TRADE SECRET ASSET MANAGEMENT 2023

A Guide to
Information Asset Management
With RICO, Blockchain, and COVID

R. Mark Halligan

and

Richard F. Weyand

ISBN 978-1-954903-13-5
Printed in the United States of America

WA

Weyand Associates, Inc.
Bloomington, Indiana, USA
January, 2023

CONTENTS

Introduction

"Trade Secret Asset Management 2023" is intended to serve company employees, owners, and attorneys as a quick course in the essential issues surrounding the identification and management of trade secret intellectual property, yet it is comprehensive enough to provide the reader with a working understanding of the nature of proprietary trade secret information and its proper stewardship. It is intended for and accessible by both attorney and non-attorney readers who need an understanding of trade secret matters for the performance of their roles within the company. This book will give the reader the vocabulary and conceptual framework required to meaningfully discuss trade secret matters with counsel.

"Trade Secret Asset Management 2023" is divided into four main sections. The first section, comprising Chapters 1 through 4, defines trade secret information, discusses the manner in which trade secrets are defended in court, and considers at some length the various ways in which trade secret property rights in information can be lost.

Chapters 5 through 11 discuss the all-important security issues surrounding trade secret information in considerable detail, including securing proprietary information against both insider and outsider threats, establishing a trade secret culture, monitoring for trade secret information in the public domain, and the often ignored issue of inbound security. Also included in this section are suspect profiles of trade secret misappropriators, both domestic and foreign.

Chapters 12 through 15 discuss accounting for trade secrets, a more recent development in the proper stewardship of proprietary information, including the inventory, identification, classification, valuation, and reporting of trade secrets, as well as trade secret life-cycle management. Chapter 15, new for this edition, discusses the trade secret audit, a necessary process in assessing the company's handling of its trade secrets.

The final section considers the future of trade secret asset management. Chapter 16 discusses the Defend Trade Secrets Act of 2016 including the drivers for its passage, the terms of the Act, and our view on its implications for trade secret management into the future. Chapter 17, new for this edition, discusses the RICO provisions in the Defend Trade Secrets Act of 2016 and their implications. Chapter 18, new for this edition, discusses the impact the changes brought about by the COVID pandemic will have on trade secret protection. Chapter 19 discusses automated trade secret asset management, its methods, benefits, and implications for the future, including the use of blockchain technology. Chapter 20 describes Trade Secret Examiner, a commercial trade secret asset management system, which has been developed along the lines described in Chapter 19.

The Appendices include various materials supportive of the text, including the texts of relevant laws and illustrative court decisions in trade secrets cases. Of special interest to many readers will be an extensive trade secret checklist for a typical manufacturing company and a sample trade secret exit interview form. New for this edition are Appendix D, a list of documents required in performance of the trade secret audit of Chapter 15, Appendix G, a list of Trade Secret Examiner® system features, and Appendix H, an analysis of the DTSA changes to the EEA by Dennis Crouch.

R. Mark Halligan
Naperville, Illinois
January 1, 2023

rmarkhalligan@fisherbroyles.com
312-607-0102

Richard F. Weyand
Bloomington, Indiana
January 1, 2023

weyand@rcn.com
630-527-0000

PART I: THE LAW

Trade secret is one of the most elusive and difficult concepts in the law to define.

— United States Court Of Appeals For The Fifth Circuit

Chapter 1
What is a Trade Secret?

Executive Summary

A company's assets are composed of both tangible and intangible assets. Intangible assets include both reputation assets and intellectual property, which includes copyrights, trademarks, patents, and trade secrets. Only patents and trade secrets protect ideas. A trade secret is information that is valuable for being not generally known in the trade and that the information holder takes reasonable measures to keep secret. Trade secrets may exist in all facets of the company's operations, and may include information that is not actively used in the company's operations as well as information about what does not work. The economic value of a trade secret may be destroyed through disclosure or a failure to take reasonable measures to protect the secrecy of the information.

Tangible and Intangible Assets

A company's assets are composed of tangible and intangible assets. Tangible assets include both physical and financial instruments owned by the company, and are accounted for under current accounting methods. Intangible assets include the reputation of the company and its products as well as its intellectual property, and are not accounted for under current accounting methods.

Tangible assets include all the physical objects the company owns — land, buildings, vehicles, furniture, office equipment, manufacturing equipment, parts inventory, product inventory. Tangible assets also include all the financial resources of the company — bank balances, stocks, bonds and security holdings, loans owed to the company, accounts receivable less allowance for non-collection, surrender value of insurance policies.

Intangible assets include the reputation of the company, known as goodwill, and the reputation of its products, known as branding. Intangible assets also include the company's intellectual property. There are four types of intellectual property: copyrights, trademarks, patents, and trade secrets.

Intellectual Property

Copyright is the statutory right of an author to control the reproduction, adaptation, distribution, performance, and display of an original work of authorship. Copyright law protects only the originality of the expression, not the underlying ideas. The expression may take the form of the written word, or it may be in the form of a drawing, picture, sculpture, or other medium. Copyright vests in the author from the moment of creation, may be registered with the Library of Congress, and is protected under federal law. The important thing to realize in our context is that copyright protects the method of expression, but not the content of the expression. In particular, with regard to ideas, the ideas expressed in a written work or drawing are not protected under copyright.

Trademark refers to any source-identifying device, such as a logo or symbol, whether used in connection with products or services. Trademark may take the form of words, names, symbols, logos, figures, letters, colors, and even smells. Trademarks may be registered with the federal government and protected under federal law, or they may remain unregistered and protected at common law. Once again, the important thing to realize in our context is that trademark protects source identifiers and not ideas.

Patent is an official certificate that describes a new invention. The inventor applies for a patent to the federal government, which performs an *ex parte* examination to determine whether the invention meets the rules for patent, including that the invention must be novel, useful, and non-obvious. The owner of the patent is entitled for a limited time to exclude any person or entity from making, using, selling, or offering to sell the invention defined in the patent, after which time the invention passes into the public domain and may be used by anyone. Patent is the first intellectual property right we have discussed that protects ideas.

Trade secret is, broadly, the right of a company to maintain internal information as a secret. Unlike copyrights, trademarks, and patents, there is no formal procedure for examination and registration of trade secrets with the government. Also unlike copyrights, trademarks, and patents, trade secrets are generally protected under state laws. Trade secrets, like patents, protect ideas, and every patent begins life as a trade secret. Trade secret protection applies to any information that is sufficiently valuable to provide the information owner an actual or potential competitive advantage in the marketplace. Trade secret protections are thus broader than patent protections, as trade secrets need not meet the patent requirements of novel, useful, and non-obvious. Also unlike patents, trade secrets need not be disclosed in an application, and do not revert to the public domain after a finite term.

Definition of a Trade Secret

Individual states have enacted different laws to protect trade secrets, though most states have adopted the Uniform Trade Secrets Act. The full text of the UTSA is contained in Appendix A.1. The theft of trade secrets is also a federal criminal offense under the Economic Espionage Act of 1996. The full text of the EEA is contained in Appendix A.3.

The UTSA includes a definition of a trade secret:

> "Trade secret" means information, including a formula, pattern, compilation, program, device, method, technique, or process, that:
>
> (i) derives independent economic value, actual or potential, from not being generally known to, and not being readily ascertainable by proper means by, other persons who can obtain economic value from its disclosure or use, and
>
> (ii) is the subject of efforts that are reasonable under the circumstances to maintain its secrecy.

The first thing to notice is the broadness of the definition: trade secret means information. Any information in the company's possession that meets the two enumerated requirements qualifies as a trade secret. Trade secret is thus much broader than patent, copyright, or trademark in its potential inclusion of any information possessed by the company.

The first requirement on the information in order to be considered a trade secret is that it "derives independent economic value … from not being generally known to … other persons." Here we have that aspect of a trade secret that makes it an asset, contained right in the definition: it must have economic value. Notice also that this economic value is dependent on the information not being known to others. Phrased another way, disclosure of the information will destroy the economic value of the trade secret, a subject we will have much more to say about later.

Note also that there is no requirement in the definition that the trade secret be actively used in the company's business. Suppose that the company discovers a new process that may be used in the manufacture of the company's products, but that another new process performs even better. Both can be trade secrets, even though the first process is not used because the second is preferred. The first process could assist competitors who are unaware of either process, and so the company derives economic benefit from those competitors not knowing about the first (unused) process as well as the second and better one.

The second requirement on the information in order to be considered a trade secret is that it "is the subject of efforts ... to maintain its secrecy." That is, the owner of the trade secret must try to keep it a secret. This effort is not optional. Failure to make such an effort will destroy the trade secret rights and the inherent economic value of the trade secret. The phrase "reasonable under the circumstances" is a matter of debate — and of litigation — in specific cases, but the necessity for security to maintain the secrecy of the trade secret is clear.

Examples of a Trade Secret

The obvious trade secrets that first come to mind for most companies are those that relate directly to the design and manufacture of the company's products. Research, development, and engineering trade secrets may include the results of laboratory tests, the design of product and manufacturing equipment prototypes, and the company's testing and evaluation processes. Manufacturing trade secrets may include the manufacturing processes, raw materials and proportions, and supplier names and contract terms.

But trade secrets exist in other areas of the company as well. Marketing trade secrets may include the results of customer or consumer surveys, plans for advertising campaigns, discount structures, and market analyses and projections. Sales trade secrets may include the structure of sales incentive plans, contact information for targeted or key customers, and customer vetting processes for sales promotions or special treatment. Financial and accounting trade secrets may include financing plans for new facilities, quarterly financial projections, and pre-release quarterly results.

Trade secrets can also include information about what does not work, called "negative know-how". If the company has tried 40 formulations for a new process, all of which failed, before succeeding with the forty-first formulation, all forty-one formulations may be trade secrets. The relevant issue here is that a competitor would economically benefit from knowing that those forty formulations did not work, saving it the time and expense of trying all of those formulations in developing its own process. If reasonable measures are taken to maintain the secrecy of these failed formulations, they qualify as a trade secret under the definition.

Finally, we have mentioned that novelty is not required for trade secret protection, and in fact trade secrets are often combinations of known art. It is well established that a protectable trade secret can exist in a combination of known elements that together afford a competitive advantage. The fundamental test is the economic value, actual or potential, derived from the secrecy of the information vis-à-vis competitors. The idea or information need not be complicated. It may be intrinsically simple and nevertheless qualify as a trade secret unless it is generally known in the trade or readily accessible from a public or well-known source in the trade.

An extensive list of potential trade secrets of a typical manufacturing company is included in Appendix 3.

Summation

Trade secrets can exist in all areas of the corporation. Trade secrets include any information that derives economic benefit from being not generally known and that is the subject of reasonable efforts to maintain secrecy. Trade secrets thus cover a much broader range of information than copyright, trademark, or patent. Trade secret information can include information about what does not work, as well as information that is not being actively used in the company's operations. Trade secret value is destroyed by general disclosure, as well as by a failure to take reasonable security measures.

Chapter 2
The Nature and Importance of Trade Secrets

Executive Summary

The differences between patents and trade secrets make one protection method or the other more appropriate depending on the nature of the information and the company's goals. Trade secrets comprise the core competencies of the corporation, and account for a large portion of the difference between a company's book value and its market capitalization. The shareholder value of a corporation is thus strongly dependent on the security and resulting value of the company's trade secrets. Theft of trade secrets is a large and growing problem, of which many companies are unaware and for which most are unprepared.

The Differences and Benefits of Patents and Trade Secrets

The distinctive nature of trade secrets can be seen by comparing trade secrets to patents. Both intellectual property rights protect ideas, but there are significant differences between the two. The primary differences are summarized in Table 2-1.

First, patents are limited to inventions — apparatus and methods that produce concrete and tangible results. Recipes, sales forecasts, financial results, vendor and customer lists, business plans, and many other types of important company information do not qualify. Trade secrets, on the other hand, can include any information held by the company.

The requirement for a patent is that the invention be novel, useful, and non-obvious. None of these requirements apply to trade secrets. The trade secret need not be novel, as some of the company's competitors may also be using the information and holding it as a trade secret as well. The trade secret need not be useful; as we have seen, knowing what does not work can also qualify as a trade secret. Finally, the trade secret need not be non-obvious in the patent sense: it can be a combination of well-known steps. The only requirements for a trade secret are that it provides the company with a competitive advantage from not being generally known and the company has taken reasonable measures to maintain its secrecy.

Table 2-1 The Differences Between Patents and Trade Secrets

	Patents	**Trade Secrets**
Coverage	Inventions	Any information
Requirements	Novel, useful, non-obvious	Competitive advantage from being secret, measures taken to maintain secrecy
Examination	By government examiner	None
Disclosure	On application	Never
Term	20 years	Indefinite
Public Domain	On expiration	Never
Costs	Application & attorney fees	Cost of maintaining secrecy
Reverse Engineering	Prohibited	Allowed
Independent Development	Prohibited	Allowed

Patents are granted by the federal government after review by a patent examiner. Before granting a patent allowance, the patent examiner will insure that the invention is patentable subject matter — an apparatus or method providing a concrete and tangible result — and that it is novel, useful and non-obvious. Other requirements, such as that the patent disclosure must be sufficiently enabling to allow one skilled in the art to reproduce the invention, will also be examined and ruled on. There is no similar examination provided by any government agency for trade secrets.

Patents must be disclosed at the time of application. This disclosure must be sufficiently detailed to allow one skilled in the art to reproduce the patent. In addition, the disclosure must disclose the best method of implementing the invention. The inventor cannot disclose part of the invention, or disclose an inferior manner of implementing it, while retaining the best way to implement the patent as a secret. There is no requirement for disclosure of a trade secret. In fact, disclosure of a trade secret can destroy trade secret rights in the information.

Patents are granted for a limited term, currently twenty years from the date of first application. Trade secrets have no expiration period, and trade secrets can be successfully held indefinitely if secrecy is maintained. Patents revert to the public domain at the expiration of their term. From that point forward, anyone can use the patent for their own purposes. Trade secrets need never revert to the public domain, and may be retained by the company indefinitely.

Additionally, there are significant application costs and attorney's fees associated with submitting and prosecuting a patent application through the U.S. Patent and Trademark Office. If worldwide protection is sought, application costs and attorney's fees increase rapidly with patent applications submitted in multiple foreign jurisdictions. There are no costs associated with a trade secret beyond making the required effort to maintain the secrecy of the information.

Finally, reverse engineering and independent development of patents are prohibited. That is, any use of the patented invention during the term of the patent is reserved to the inventor and his licensees. Trade secrets provide no such protections. If the information can be discerned upon inspection or reverse engineering of the product, the product itself constitutes a disclosure and the information can be used by the reverse engineer. Also, trade secrets can be legally developed independently and used. They may in turn be held as a trade secret by the independent developer or disclosed at his option.

These differences underscore the nature and importance of trade secrets to the company. Most information developed by the company will not qualify for patent protection, but the same information will often qualify for trade secret protection. Patents are publicly disclosed in the application process, and revert to the public domain on the expiration of the twenty-year term, while a trade secret can remain secret forever. In addition, even the core trade secrets of many companies would be cost-prohibitive to protect with worldwide patents due to the costs of the patent system.

At the same time, if the information can be discerned or reverse engineered from the product, the information cannot be held as a trade secret. Further, if the information is likely to be independently developed, holding the information as a trade secret will prove difficult.

Table 2-2 presents the decision parameters for determining whether to apply for a patent or hold information as a trade secret.

Table 2-2 Apply for Patent Or Hold as Trade Secret?

Trade Secret *if* Information is not patentable subject matter

Trade Secret *if* Information is not novel, useful, and non-obvious

Trade Secret *if* Disclosure is not desired

Trade Secret *if* Longer than 20-year term is desired

Trade Secret *if* Patent application and attorney's fees are prohibitive

Patent *if* Reverse engineering from the product is possible

Patent *if* Independent development is likely

Core Competencies Are Embodied in Trade Secrets

At one time, corporate strategies often included vertical integration, the ownership and control of the supply and distribution chains. Over time, it became clear that a better corporate strategy is to concentrate on the company's core competencies, those areas in which the company has exceptional abilities that present the opportunity for competitive advantage in the marketplace. These core competencies result from the company's unique knowledge of how to perform the key tasks required in meeting marketplace demand in their business area.

That is, the company's core competencies are the direct result of the company's trade secrets.

Consider again the definition of a trade secret from Chapter 1: "trade secret means information … that derives independent economic value … from not being generally known to … other persons." Companies possess trade secret information that gives them a competitive advantage over 'other persons' in their market, and constitutes their core competencies. We see now why the vertical integration strategy failed: the companies did not possess the trade secret information required to run those new business operations they acquired up and down the supply and distribution chains, and they were at a disadvantage compared to established companies that did.

The vertical integration strategy failed not because the economies of vertical integration did not theoretically exist, but because these economies were overwhelmed by the large competitive advantage conferred by trade secrets.

Book Value vs. Market Capitalization

To understand the rising value of trade secrets to the modern company, one need only compare the accounting book value of the company to its market capitalization, the value of its stock. Prior to 1960, the overwhelming majority of the shareholder value of most companies was contained in their physical assets, as accounted for in their book value. The land, buildings, equipment, and vehicles of old economy companies dominated their valuations.

In today's information economy, the book value of many companies approaches zero. High technology firms with leased buildings, leased furniture, leased computer equipment, and no physical plant, manufacturing equipment, or service vehicles nevertheless see their stock trading for hundreds of dollars per share and market capitalizations in the tens of billions of dollars. In the information economy, it is necessarily the value of information that drives shareholder value.

The Brookings Institution's Margaret Blair reports, "Depending on the measure used, at least 50%, and possibly as much as 85%, of the assets and other sources of value in the corporate sector do not appear on the books of corporations. In some firms, the gap between so-called "book" value (what the accountants say the assets of the firm are worth) and the firm's market value is modest, but in others the gap is as much as 95%." She continues, "The information needs of the new economy have already outgrown what traditional accounting systems can deliver."

The difference between book value and market capitalization is the value the markets place on a company's intangible assets, those assets that are not accounted for when calculating book value using traditional accounting methods. The vast bulk of the value of those intangible assets is comprised of the company's trade secrets, not of its goodwill, branding, or other intangible assets. Trade secrets are what allowed a Google to come out of nowhere to dominate the search engine business over competitive search technologies from companies with established goodwill and branding like Yahoo, AOL, and Microsoft. It is the trade secrets that drove their success, which in turn drove their goodwill and branding, not the other way around.

Google's book value at the end of Q2 2015 was $119 billion, of which $84 billion was in cash, marketable securities, and receivables, while its total assets in property, plant, and equipment were just $27 billion. Google's market capitalization at that same point was $360 billion, over 13 times the value of its physical assets. Over $240 billion of that market capitalization is completely unaccounted for by traditional methods. A quarter of a trillion dollars unaccounted for.

This dominance of trade secret information in the value of companies has its dark side as well. If a company's trade secrets are lost or compromised, it is possible for the company's stock value to plummet. There is no safety net in the book value of information technology companies to protect shareholder value when trade secret information is lost or compromised. Google's book value at the end of Q2 2015 amounted to $174 per share, while its stock traded at $525 per share.

A new and better technology that obsoletes a company's trade secrets, or a market re-evaluation of the economic advantage provided by the company's trade secrets, can also result in severe and sudden drops in shareholder value. Information economy companies cannot afford to stand still and allow their trade secrets to be obsoleted by leaner and meaner competitors. Indeed, they must do it themselves, in an internalization of the creative destruction of the marketplace. The result is that the trade secret portfolio of an information economy company is necessarily a turbulent, dynamic environment as trade secret information is discovered, used, and ultimately discarded as it is obsoleted by new trade secrets invented within the company.

Likelihood of Theft of Trade Secrets

If a company's trade secret information is such a large part of its shareholder value, and the loss or compromise of a company's trade secret information can result in catastrophic shareholder losses, the obvious question to ask is, How likely is such a loss or compromise? The answer is sobering.

The American Society of Industrial Security (ASIS), PricewaterhouseCoopers and the U.S. Chamber of Commerce conduct a biannual survey whose 2002 edition estimated that between $53 and $59 billion in proprietary information was lost by the largest U.S. companies between July 2000 and June 2001. These are the reported losses of which companies were aware and to which they admitted in the survey. As most of the respondent companies have no formal reporting mechanism for trade secret theft, the numbers can be assumed to be understated. In a speech in 2003, FBI Director Robert S. Mueller III claimed that as much as $200 billion annually is lost to economic espionage.

The Annual Report to Congress on Foreign Economic Collection and Industrial Espionage—2004, published by the Office of the National Counterintelligence Executive, reported, "Foreign individuals from both the private and public sectors in almost 100 countries attempted to acquire sensitive US technologies in fiscal year 2004...." U.S. companies often seem blissfully unaware of the danger, as the ONCIX report noted, "Most of the foreign entities attempting to acquire US technology last year employed tools and techniques that were easy to use, inexpensive, low risk, and sometimes legal. In a majority of cases, foreign collectors simply asked—via e-mail, phone call, FAX, letter, or in person—for the information." As for whether the situation is likely to improve, the ONCIX report noted, "The CI Community expects no decline in foreign demand for sensitive US technologies over the next few years. The United States remains the source of much of the world's most advanced technology, and, in many industries, foreign entities depend on that innovation to improve their competitiveness."

While the focus of the Office of the National Counterintelligence Executive is on economic espionage by foreign entities, U.S. companies reported in the ASIS survey that their primary risk factor for proprietary information loss is former employees. The complete ranked list of risk factors from the ASIS report is given in Table 2-3. Former employees, current employees, and on-site contractors are particularly troubling as security risks. The company cannot lock up its trade secrets from employees if the trade secrets are to be used in the company's operations to provide the competitive advantage they make possible. Use of the trade secret information requires disclosure to employees. We will have much more to say about this topic later.

Table 2-3 Risk Factors Associated With Proprietary Information and Intellectual Property Loss

Former Employees
Foreign Competitors
On-Site Contractors
Domestic Competitors
Computer Hackers
Vendors/Suppliers
Current Employees
Strategic Partners
Intelligence Services
OEMs/Outsource Manufacturers
Media

What is clear is that the loss of trade secret information is a very real and on-going problem, and puts the information-driven shareholder value of the corporation at risk. The odds are good that the trade secret information of your company is being stolen as you read this book.

Summation

We have seen how the differences between patents and trade secrets derive from their legal definitions, and how those differences make one protection method or the other more appropriate depending on the nature of the information and the company's goals. The company's trade secrets comprise and encompass its core competencies, and are the critical knowledge that provides the opportunity for competitive advantage. The trade secrets of the company account for much of the difference between the company's book value and its stock market capitalization, which makes shareholder value strongly dependent on the security and value of the company's trade secrets. The company's trade secrets are at risk of theft, this risk is large and growing, and most companies are insufficiently aware of or prepared to deal with this risk.

Chapter 3
How Trade Secrets Are Defended

Executive Summary

Trade secrets can only be validated in litigation. To successfully assert trade secret property rights, plaintiff must make four proofs: existence, ownership, notice, and access. Existence is proved according to the legal definition of a trade secret, guided by a consideration of the six factors of a trade secret from the Restatement (First) of Torts. Ownership usually requires a trade secret assignment agreement executed by the creator of the information. Notice that a company considers the information a trade secret must have been provided. Access to the information must be shown in order to overcome a defense of independent development or reverse engineering.

Litigation and the EONA Proofs

Given that trade secrets are so valuable and important to the company and to shareholder value, we must address the issue of how trade secrets are defended. Specifically, how are the legal protections for trade secret property rights invoked?

The only way to validate a trade secret is through litigation. Absent a court finding that trade secret property rights in information exist, the trade secret status of the information remains alleged but unproven. This contrasts with the situation for patents, in which an *ex parte* examination has resulted in an official government certificate and a presumption that the patent property rights are valid. Without such a formal certification process, trade secret property rights are not validated in a legal sense until they are litigated.

Litigation of trade secret rights occurs after an alleged misappropriation of the information. The trade secret owner files suit as plaintiff that the defendant violated plaintiff's trade secret property rights in the information. Plaintiff must make six proofs in order to prevail in the litigation. These proofs are conjunctive; that is, they must all be proven. Failure on any of the six proofs will result in failure of the suit. The six proofs are detailed in Table 3-1.

Table 3-1 Six Proofs Required to Prevail as Plaintiff in Alleging Trade Secret Misappropriation

Existence: The information qualifies as a trade secret; i.e. a trade secret exists.

Ownership: Plaintiff has ownership rights in the information.

Notice: Defendant had actual, constructive, or implied notice of the trade secret status of the information.

Access: Defendant had access to the information; i.e. defendant did not independently develop the information.

Use: Defendant has used or disclosed the information, or the use or disclosure of the information is threatened.

Damages: Plaintiff was harmed by defendant's use of the information, or defendant was unjustly enriched thereby.

Of these six proofs, the first four, the so-called EONA proofs, are unique to trade secrets. In contrast, for patents the first four proofs are presumed. The *ex parte* process of patent application and examination certifies the existence and ownership requirements of the first two proofs, while publication of the patent creates the notice and access conditions of the next two proofs. In patent litigation, therefore, only use — making, using, selling, or offering to sell the invention — and damages need to be proved.

The EONA proofs of a trade secret lawsuit — existence, ownership, notice, and access — substitute for the application, examination, and publication steps of the patent process. Since there has been no formal application process in advance, the trade secret plaintiff must present the EONA proofs to validate those trade secret property rights during litigation.

It is important to note that the ability to succeed at each of the EONA proofs depends on actions the information owner takes before the misappropriation occurs. It is usually too late at the time of litigation to correct deficiencies in the stewardship of information assets prior to the misappropriation of these assets. This will be described in more detail as we discuss each proof in turn.

The last two proofs required during trade secret litigation — use and damages — are similar in substance to the use and damages proofs required during patent litigation, and will not be discussed further here. Instead the emphasis will be on the validation of trade secret rights during litigation with the EONA proofs. The EONA requirements for a successful assertion of trade secret property rights in litigation provide the basic foundation for trade secret asset management.

Existence I: The UTSA Definition

The existence proof requires the plaintiff in a trade secret misappropriation lawsuit to show that the information meets the legal requirements for trade secret status and protection. The legal requirements for trade secret protection are spelled out in the definition of a trade secret in the Uniform Trade Secret Act (UTSA) , contained in Appendix A.1. In practice, this is comprised of four portions:

> 1) "Trade secret means information ... that ... derives independent economic value, actual or potential"
>
> 2) "from not being generally known to [other persons who can obtain economic value from its disclosure or use]"
>
> 3) "and not being readily ascertainable by proper means by other persons who can obtain economic value from its disclosure or use, and"
>
> 4) "is the subject of efforts that are reasonable under the circumstances to maintain its secrecy."

Plaintiff must first show that the information has economic value. This value may be actual, in that the trade secret is in current use within plaintiff's business operations, or potential, in that the trade secret may be of value either to plaintiff, defendant, or others in the future. Further, that value must be derived from the information not being generally known to others in the trade. That is, the economic value of the information to plaintiff must be reduced by the disclosure of the information to the defendant or others, or the defendant or others must be unjustly enriched by the disclosure of the information. Defendant will argue that the alleged trade secret information has no economic value, and that what value it might have is not derived from the secrecy of the information, and so trade secret status is inappropriate.

Second, plaintiff must show that the information is not generally known to others who could make use of it. Defendant will argue that the information is widely known, and that trade secret status is therefore inappropriate. It is important to note here that defendant must make an affirmative showing that the information is generally known in the trade to prevail in this argument. It is not enough to simply assert that "Everyone knows that." Absent evidence or expert testimony that the information is generally known in the trade, plaintiff's assertion that the information is not generally known will prevail. Put another way, if it is generally known in the trade, defendant should be able to easily produce evidence that this is the case.

Third, plaintiff must show that the information is not readily ascertainable by proper means. That is, that the alleged trade secret cannot be readily reverse engineered or independently developed without access to plaintiff's information. Defendant will argue that the information can be independently developed with little effort and without access to plaintiff's alleged trade secret information.

Finally, and most importantly of all, plaintiff must show that he took measures "reasonable under the circumstances" to maintain the secrecy of the information. This is in fact where trade secret lawsuits are most likely to fail. The courts will not grant trade secret protection after the fact to information that the information owner did not take reasonable measures to protect in the first instance. Lapses in security measures, failure to treat the information as a secret, and inadvertent disclosure of the information outside the bounds of a confidential relationship are each enough to destroy the potential trade secret status of the information. Defendant will require disclosure of plaintiff's security measures, and argue at trial that these measures were insufficient to protect the information assets as trade secrets.

It is possible to forfeit the company's entire portfolio of trade secret assets with a failure of the existence proof. This exact scenario came to pass in *Omega v. Chroma*. Several former employees of Omega left the company and set up a competitive firm using technical, manufacturing, and sales information from Omega. The former Omega employees among them had complete knowledge of Omega's business operations. Omega sued Chroma for trade secret misappropriation. The state court found that, while the information was protectable as trade secret information and that defendants had in fact used the information, Omega had failed to provide notice and take measures reasonable under the circumstances to maintain the secrecy of the information. The information thus did not qualify for trade secret protection under the law, and Chroma was free to use the information. The lower court was sustained by the Vermont Supreme Court, and Omega lost its entire portfolio of trade secrets. The entire text of the *Omega v. Chroma* decision is contained in Appendix B.2.

It should also be noted that the courts do not apply a single standard of security measures for all companies and all circumstances. What is reasonable under the circumstances will change when the circumstances change. If a computer hacking incident involves the theft of trade secret information, reasonable security measures must subsequently include protections against this known threat. Mom-and-pop companies will not be held to the same standards of security as large corporations that have more staff and resources to implement more complex procedures. Finally, the intrinsic value of the trade secret information itself will have an impact on the expected extent of security measures necessary to meet the standard of reasonableness.

Existence II: The Six Factors from the Restatement (First) of Torts

The Uniform Trade Secrets Act is based on the modern definition of a trade secret. Trade secret is a concept with a long history in the common law, however, and the considerations appropriate for determining the existence of a trade secret in a misappropriation suit are much older than the UTSA. The Restatement (First) of Torts in 1939 provided six factors to be considered in determining whether a trade secret exists, and these six factors are considered in every trade secret misappropriation suit. The six factors are detailed in Table 3-2. We will discuss each of the six factors in turn.

Table 3-2 Six Factors of a Trade Secret from the Restatement (First) of Torts

"An exact definition of a trade secret is not possible. Some factors to be considered in determining whether given information is one's trade secret are:

(1) the extent to which the information is known outside of his business;
(2) the extent to which it is known by employees and others involved in his business;
(3) the extent of measures taken by him to guard the secrecy of the information;
(4) the value of the information to him and to his competitors;
(5) the amount of effort or money expended by him in developing the information;

(6) the ease or difficulty with which the information could be properly acquired or duplicated by others."

"The extent to which the information is known outside of his business" captures the "not generally known" clause from the UTSA definition. Obviously, if the information is widely known outside of the business, the argument for trade secret status is weaker. But the test in the six factors is extent: this is not a yes or no test. If some of the competitors know the information and hold it as a trade secret, and others don't, there is still an economic advantage accruing to the company with knowledge of the trade secret. Further, even if all competitors know the information and treat it as a trade secret, but the information is not generally known, there is still an economic advantage to the company in that the lack of the information is a barrier to entry to potential new competitors entering the field. Of course, the optimal situation for this factor is no knowledge of the information by anyone outside the company.

"The extent to which it is known by employees and others involved in his business" speaks to another issue we will discuss in the chapters on security: "need to know". This factor considers the extent of secrecy within the company. It is based upon the observation that the more employees there are who know the information within the company, the less likely it is that the information is a trade secret. Disclosure of valuable information should therefore be limited to those who have a need to know the information in order to perform their work. Every secretary, mailroom clerk, and janitor in the company does not need to know the secret formula for the company's product. The more restricted the information is within the company, the more likely that the information qualifies as a trade secret.

"The extent of measures taken by him to guard the secrecy of the information" corresponds to the requirement to take "reasonable measures to maintain the secrecy of the information" in the UTSA definition. Once again, the security measures taken by the information owner are of critical importance in determining whether trade secret property rights in the information are appropriate. Where the UTSA requires that measures be "reasonable under the circumstances", the six factors of the Restatement again provide a sliding scale rather than a yes or no test. Information is more likely to qualify for trade secret protection if the owner demonstrates due care and proper stewardship to protect the information.

"The value of the information to him and to his competitors" echoes the "derives independent economic value" requirement of the UTSA. Trade secrets must have independent economic value derived from the secrecy of the information. The modern law of trade secrets does not require that the information owner actually use the information, but there must be actual or potential economic value derived from the secrecy of the information. The economic value could lie in the fact that competitors, absent the information, must spend additional time, effort, or money. This is particularly true in the case of negative know-how, where the uninformed competitor will have to repeat all the negative trials and go down all the blind alleys that the information owner already knows will not or cannot work.

"The amount of effort or money expended by him in developing the information" is an expression of the economic value of the information. Trade secret information derived at great cost to the information owner is given greater consideration than information that was obtained with less time, effort, and money. Consider the case we discussed before, where two different methods exist for a process. Both can be trade secrets, even if the earlier method has been supplanted by a later and better one. The first method could have been discovered at great cost, while the second could have resulted from a flash of insight. The cost incurred in uncovering the first method will give it greater weight in consideration of this factor.

"The ease or difficulty with which the information could be properly acquired or duplicated by others" is the analog of the UTSA definition's "not being readily ascertainable by proper means" clause. The company's competitors have the right to independently develop or reverse engineer the company's trade secret information. The extent to which such independent development or reverse engineering is possible serve to lessen the weight given to trade secret status under this factor.

It should be noted that meeting all of the six factors is not required. However, all six factors will be considered at trial, and the probability of the existence or non-existence of a trade secret can only be determined after all the six factors have been evaluated and considered. An excellent discussion of the six factors and other issues discussed in this chapter is contained in *Learning Curve Toys, Inc. v. Playwood Toys, Inc.*, a decision that captures the modern view of trade secret law in the United States. The full text of *Learning Curve Toys, Inc. v. Playwood Toys, Inc.* is contained in Appendix B.4.

Ownership

Ownership is often overlooked in trade secret litigation, but it does come up in specific cases. Plaintiff must show that he owns the information alleged to be a trade secret. There are traps with regard to the ownership of trade secret rights that can ensnare the unwary.

The "work made for hire" doctrine seems to suggest that ownership rights in works created by an employee vest in the employer. However, only the United States Copyright Act contains a statutory "work made for hire" provision. There is no "work for hire" provision in the UTSA. Except in the narrow circumstance where an employee is hired specifically to invent something, the common law does not automatically assign trade secret ownership rights to the employer absent an express assignment of invention and trade secret rights in an employment agreement.

The written assignment requirement also applies to independent contractors, outside manufacturers, and other suppliers to the company. Trade secret assignments need to be executed to transfer ownership rights from the creator to the company. If the work was done on company time, on company property, or with company equipment, the company may obtain a royalty-free "shop right" to use the trade secret, but ownership of the trade secret remains with the person or persons, whether employee or outsider, who created or discovered the trade secret information.

The ownership proof does not come up often in litigation, but when it does, it is devastating. The employee, contractor, or vendor who never gets around to signing the employee trade secret assignment agreement, and the harried human resources or purchasing staff that fails to follow up on the requirement, create a situation in which the ownership rights in trade secret information created by the employee, contractor, or vendor remain in the possession, custody, and control of the creator even if the company paid for the work.

This situation can also arise where a contract employee is hired through an agency. Even though the agency has assignment agreements on file with the company, if the agency has not executed similar assignments with the contract employee, that contract employee is not bound by the agency's agreement with the company. Such contracts can also be deemed at law to be non-transferable "personal" contracts, requiring new trade secret assignment agreements to be executed with the new entity upon any merger or acquisition of the vendor or contract agency by another firm. For these reasons, individual trade secret assignment agreements should be obtained directly from all employees, contractors, outside manufacturers, and suppliers of the company, and renewed on every extension, renewal, or transfer of contracts.

Notice

Employees are free to use their general skills, knowledge, and experience in competition with their former employers. However, former employees are not free to use or disclose the trade secrets of their former employers. The notice proof requires the plaintiff to demonstrate that plaintiff drew the dividing line between the employee's general knowledge, skills, and experience and the company's alleged trade secrets. That is, the plaintiff must show actual, constructive, or implied notice of the trade secret status of the information with regard to each defendant accused of misappropriation.

Note that it is not sufficient to claim that "Everything we do is a trade secret." The courts have taken judicial notice of the fact that every company possesses some information that does not qualify as a trade secret. Failure to distinguish between trade secret information and non-proprietary information in the possession of the company thus puts all company information in a single non-protected class, and will be fatal to the plaintiff's case in litigation.

Notice can be actual, constructive, or implied. Actual notice includes labeling documents containing trade secret information with "Confidential" or "Proprietary" stamps or legends. Actual notice can also take the form of an inventory of trade secrets presented to an employee during employment or during an exit interview.

If the trade secret is highly valuable, the courts will sometimes impose constructive notice. For example, is someone loans you a diamond ring, you do not need to be told that it is valuable and needs to be secured.

Notice may also be implied from the totality of the circumstances relating to its disclosure and use within the company. For example, if all of a restaurant company's recipes are kept not on computer but in hard-copy documents that are locked in the company's safe and can be checked out only for specific purposes, the employee has been given implied notice that these are company trade secrets.

The court test at trial will be whether a reasonable person knew or had reason to know that the information was a trade secret. Best practice is to provide actual notice and remove all doubt, both in the employee's mind and at trial, that the company considers the information a trade secret.

Access

Since people have the right to independently develop or reverse engineer trade secret information, plaintiff must show that such independent development or reverse engineering did not take place, that instead the defendant came into possession of the alleged trade secret through his confidential access to the plaintiff's information. The access proof can be difficult if access to trade secret information was not controlled and access tracking procedures were not in place.

In specific cases, access may be easy to prove. The defendant may either have written or been on the distribution list for product specifications, business plans, or marketing forecasts that contained the trade secret information. The defendant may have been present, and appear in the minutes as being present, at meetings where the trade secret information was discussed. The defendant may have specific access to the information in order to perform certain job tasks. Access to the recipe, for example, will be presumed as a matter of law when the defendant's assigned job task was to mix the ingredients.

In other cases, proving access can be much more difficult. Did the defendant hear the trade secret information discussed by others in an informal setting, such as over lunch in the company cafeteria? Did the defendant date another employee who discussed the trade secret information during their off-work time together? Did the defendant pick up or read a sensitive document to which he should not have had access, and which was carelessly left out in the open when someone left their desk for a moment? Is the defendant an outsider who may have obtained access to the information through some illegal act, such as computer hacking?

In practice, proximity to the trade secret information, such as working in the same facility where the information is developed or used, together with a lack of evidence produced by the defendant showing actual independent development efforts or independent reverse engineering activities, will result in an inference of unauthorized acquisition. Access controls and safeguards, together with access tracking, will provide an affirmative proof that defendant had access to the trade secret information. This is much preferred to relying on a constructive inference of access at trial.

Summation

Trade secret property rights can only be validated in litigation. Plaintiff in a trade secret misappropriation suit must prove not only use and damages, but existence, ownership, notice, and access. Proving existence necessarily involves the legal definition of a trade secret, guided by a consideration of the six factors of a trade secret from the Restatement (First) of Torts. Plaintiff must prove he had ownership rights in the information alleged to be a trade secret, that the defendant was under notice that the company considered the information a trade secret, and that the defendant had access to the plaintiff's trade secret information.

Chapter 4
How Trade Secrets Are Lost

Executive Summary

As all four of the EONA proofs are required to successfully assert trade secret property rights in information, failure of any of the proofs will result in the loss or forfeiture of trade secret rights. Failure of the existence proof is common, and its requirement that the company take reasonable measures to maintain the secrecy of the information is the most common way to forfeit trade secret rights. Inadvertent and careless disclosures also destroy trade secret property rights in information, as do unprotected disclosures to third parties. Failure to take reasonable measures can result in the loss of trade secret property rights even if the actual access to the information was proper and not enabled by any lapse of security.

The Principle of Free and Open Information

Under U.S. law, all information is in the public domain, free and open to be used by anyone for any legal purpose, with few exceptions. These exceptions include personal credit, health, financial, and other information protected under various state and federal privacy laws such as the Fair Credit Reporting Act (FCRA), the Health Insurance Portability and Accountability Act (HIPAA), and the Graham-Bliley Act (GBA), government secrets protected under the Espionage Act and other federal laws, patents protected under the Patent Act, and trade secrets protected under the federal Economic Espionage Act of 1996 (EEA) and state statutes modeled after the Uniform Trade Secrets Act (UTSA).

The important point to be recognized here in the context of business information is that, absent trade secret protection, all unpatented business information in the possession of the company can be legally copied, disclosed, and used by anyone, including ex-employees, contractors, consultants, vendors, and competitors.

The easiest and most common way to forfeit intellectual property rights in trade secret information is failure to ensure that the information qualifies as a trade secret under the law. The resulting forfeiture leaves the information free and open for use by all, without any recourse for the company. This is in fact what happened to Omega in the *Omega v. Chroma* trade secret misappropriation suit mentioned in the last chapter and contained in Appendix B.2.

Plaintiff must successfully make all four EONA proofs to prevail in an assertion of trade secret property rights. Actions that undermine or destroy the ability to successfully make one or more of the proofs in a later litigation will put the trade secret intellectual property rights in the information at risk of forfeiture. We will look once again at the EONA proofs and consider how each might be undermined, leading to the loss of trade secret property rights in the company's information.

Failure of the Existence Proof I: The Definition of a Trade Secret

In our consideration of the existence proof we noted that the requirements of the UTSA definition of a trade secret must be satisfied to successfully assert trade secret property rights and also that the six factors of a trade secret from the Restatement (First) of Torts will be considered at trial. Failure of the existence proof then can result from a failure to meet one of the requirements of the UTSA definition, or it can result from a poor evaluation of the trade secret under the six factors from the Restatement.

The UTSA requirements, once again, are as follows.

1) "Trade secret means information ... that ... derives independent economic value, actual or potential"

2) "from not being generally known to [other persons who can obtain economic value from its disclosure or use]"

3) "and not being readily ascertainable by proper means by other persons who can obtain economic value from its disclosure or use, and"

4) "is the subject of efforts that are reasonable under the circumstances to maintain its secrecy."

The application of the first aspect of the UTSA definition to the information owner's alleged trade secret will be undermined by plaintiff's inability to prove that independent economic value derives from the secrecy of the information. Is the trade secret being used in the company's business, or are there plans to use it in the company's business in the future? If not, has consideration been given as to why the information has independent economic value to the company or its competitors? To what products or product lines does the trade secret apply? How much of the company's revenue is derived from those products? How much does that revenue depend on the continued secrecy of the information? Contemporaneous documentation of these facts, recorded before the misappropriation takes place and litigation is contemplated, will serve to establish this proof. Such documentation will be much stronger evidence at trial than after-the-fact justifications.

The application of the second aspect of the UTSA definition to the information owner's alleged trade secret will be undermined by plaintiff's inability to prove that the information was not generally known in the trade prior to the misappropriation. Is the information by its very nature not generally known outside the company, such as for the company's financial results, business forecasts, and marketing plans? For other kinds of information, were prior art searches performed and recorded? Is the company sure that the information was developed internally and not brought into the company? Have the contemporaneous records pertaining to the development of the information been retained? Again, contemporaneous documentation will be much stronger evidence at trial than after-the-fact justifications.

The application of the third aspect of the UTSA definition to the information owner's alleged trade secret will be undermined by plaintiff's inability to prove that the information was not readily ascertainable by proper means. One proper means of development that the company can prove is the manner in which the company itself researched and developed the information. Failure of defendant at trial to show evidence of a similar development effort will weaken the defense of an alleged independent development. Has the company documented the manner in which the trade secret information was discovered, refined, and developed? Have the contemporaneous records pertaining to the research and development of the information been retained? Have the costs associated with the company's research and development of the information been documented? Once again, contemporaneous documentation will be much stronger evidence at trial than after-the-fact justifications.

The application of the fourth aspect of the UTSA definition

to the information owner's alleged trade secret will be undermined by plaintiff's inability to prove that efforts to maintain the secrecy of the information were reasonable under the circumstances. Failing to take reasonable measures is the most common way in which trade secret property rights in information are lost. Even if the defendant is a former employee to whom the trade secrets were disclosed to perform his job duties, and the security measures in place have no relation to that necessary disclosure, defendant will argue that inadequate security measures disqualify the information for trade secret protection under the UTSA definition, leaving it free for use by anyone, including the defendant. The security topic is so important that we will devote the next several chapters to it.

For right now, consider the following questions. What are the most obvious ways in which sensitive information can be gathered from outside the company? Does the company implement a badge or other identification procedure, and strictly enforce these identification procedures for employees and guests? Are guests escorted at all times when on company premises? Are cameras, including those in cell phones, watches, and other electronic devices, banned from the company's facilities? Does the company have document disposal bins and on-site shredding of documents? Are passwords required for access to all of the company's computers, and is there a process in place to ensure that they are regularly changed? Are employees forbidden to use outside computers such as their own laptops or home computers for handling the company's proprietary information? All of these questions will come up in any litigation where the company asserts trade secret property rights. They are much better asked and answered — and documented — prior to any misappropriation.

Failure of the Existence Proof II: The Six Factors

While the four portions of the UTSA definition must be decided, the six factors from the Restatement (First) of Torts will be considered in making this decision. Let us consider each of the six factors in turn to examine further how the existence proof might fail. We will also consider how the company might strengthen the six-factor analysis of existing trade secrets by taking actions to increase the information's compliance with the six factors.

The first of the six factors considers the extent to which the information is known outside of the company. This is the same consideration as previously discussed for the UTSA definition, and the same questions apply. The trade secret's compliance with this factor cannot be strengthened by the company, except by ensuring that the trade secret is not disclosed outside the company.

The second of the six factors considers the extent to which the information is known by employees and others involved in the company's business. This consideration falls under the security measures portion of the UTSA definition. If everyone in the company knows the information, the security measures are not as strong as they could be. With regard to this factor, companies are often their own worst enemy. Giving the R&D group an award for coming up with a new manufacturing process is a great idea, for instance, but announcing the essence of that secret process at the company meeting where the award is given weakens the information's compliance with this factor.

Some companies unwittingly defeat the information's evaluation under this factor in the furtherance of ISO 9000 goals. ISO 9000 requires that the distribution of documents be controlled and recorded in such a way as to allow them to be updated completely, so that no employee relies mistakenly on an old version and thereby creates a quality control problem. Some companies meet this requirement by publishing all documents on the company's intranet, so that the current and correct version is always the intranet version. So far, so good. But giving every employee in the company access to those documents weakens the information's compliance with this factor.

The deference given to the company's trade secrets under this factor can be increased by instituting access controls for trade secret information, so that the only employees who have access to a given trade secret are the ones that need to know the information to perform their job duties. Such "need-to-know" access restrictions maximize the impact of this factor in consideration of the company's claim of trade secret property rights.

The third of the six factors considers the extent of measures taken by the company to guard the secrecy of the information. This is the reasonable measures requirement of the UTSA definition, and will be discussed in considerable detail in the next several chapters. Tightening security over the company's trade secrets will increase the deference given to the trade secrets under this factor.

The fourth of the six factors considers the value of the information to the company and to its competitors. This is the same consideration previously discussed for the UTSA definition, and the same questions apply. The information's compliance with this factor cannot be strengthened by the company, except by ensuring that the value of the trade secret to the company is properly documented.

The fifth of the six factors considers the amount of effort or money expended by the company in developing the information. This is another aspect of the economic value requirement of the UTSA definition. The trade secret's compliance with this factor cannot be strengthened by the company, but proper documentation of the costs expended in the development of the company's trade secret information will provide the evidence required to maximize the court's evaluation of this factor.

The sixth factor considers the ease or difficulty with which the information could be properly acquired or duplicated by others. This is the "readily ascertainable by proper means" consideration from the UTSA definition, and the same questions apply. The information's compliance with this factor cannot be strengthened by the company, except by ensuring that the company's own efforts to develop the trade secret information is properly documented.

It should be emphasized here that contemporaneous documentation for the issues under consideration in the six factors, as for the UTSA definition, will be much stronger evidence in an assertion of trade secret property rights in the information than after-the-fact justifications.

Failure of the Ownership Proof

Failure of the ownership proof is almost always the result of a failure of execution. All companies should know by now that trade secret assignments from employees, contractors, consultants, vendors, and outsource manufacturers are required. The ownership proof usually fails only when the required agreement has not been executed, and no follow-up was performed to check that such agreements are in place for all employees and third parties performing work for the company. Employee agreements with assignment provisions should be updated and renewed annually. Performance review time is a good time for this renewal, when the employee reflects on his past year's performance, sets goals for the next year, and reaps the rewards, in terms of a salary increase, of his efforts on behalf of the company. Similarly, for third-party suppliers and vendors, a good time for trade secret assignment renewals is contract renewal time. Such a trade secret assignment can be included on the reverse of the company's purchase order forms, and included as part of the terms and conditions agreed to by acceptance of the purchase order.

Failure of the Notice Proof

Failure of the notice proof is usually the result of the company's failure to conduct trade secret audits. Lacking an inventory of what the trade secrets are, the company is unable to inform employees and others to whom the trade secret is disclosed in confidence that the company is treating the information as a trade secret. Without an inventory of the trade secrets, the company is unable to ensure that all documents containing trade secrets have been marked, providing actual notice, or that all the documents containing trade secrets are properly handled, providing implied notice.

Inventorying trade secrets can be a daunting task, especially for large companies with a large number of trade secrets or companies in high-tech fields where trade secrets are created and destroyed daily in the normal course of business. Computerized methods will increasingly be applied to simplify this task. The audit and inventory of trade secrets will be discussed at length in two chapters on accounting methods for trade secrets later in this book.

Failure to provide notice creates problems for both employees and the company. Employees have the right to use their general knowledge, skills, and experience in further employment, while the company retains the rights to its trade secret information. Failure to provide notice of what information the company considers a trade secret — failure to clearly draw that line — creates a vast, fuzzy middle ground where employees are not sure what they can and cannot use in their further employment. If a former employee draws the line too narrowly, he short-changes himself and limits the benefit of his general knowledge, skills, and experience to himself and his future employers. If a former employee draws the line liberally, there is a risk of being named a defendant in a trade secret misappropriation suit brought by his former employer.

If the company refuses to draw that line between its proprietary information and the employees' general knowledge, skills, and experience, then the court will draw it at trial. The courts give deference to an employee's right to earn a living and to pursue employment with other companies, however, and the law requires the company to identify its trade secrets and give notice to its employees. As a result, if the court needs to draw that line at trial, it will protect the employees' rights by drawing the line much closer to the company than the company could justifiably have drawn it before the misappropriation. Trade secret rights in the information beyond the court's line will be forfeited by the company for failure to identify its trade secrets and provide proper notice to employees.

Failure of the Access Proof

Failure of the access proof usually results from a failure of the company to control the internal distribution of trade secret information. That is, failure of the company to implement access controls or access tracking makes it impossible to prove that the defendant had access to the information. In the typical case, a former employee goes to work for a competitor and the company subsequently becomes aware that the competitor suddenly seems to know some of the company's trade secrets. This awareness can come from the competitor's introduction of a new product, the competitor's strategic business moves or tactical sales and marketing moves, or other public information about the competitor's activities and plans. But how did the former employee get access to that information? He may not have been employed in the department or at the facility where the trade secret information was developed or used.

Lack of access controls often results in trade secret information being exposed to employees who have no obvious connection to the development or use of the trade secret information. Lack of access tracking will make it difficult to determine when the employee's activities within the company crossed paths with the trade secret information. Either deficiency will make it very difficult to reconstruct after the fact just how the employee had access to the information and to sustain the necessary access proof at trial.

Absent the company's ability to prove access, the former employee will deny access and his new employer will claim independent development of the information. These claims may in fact be true. Who's to know, when the former employee's access to the information within the company cannot be proved or disproved? Even exhaustive and expensive discovery may leave it an open question, where simple measures up front would have made the determination easy. The trade secret owner's suit will fail due to the inability to prove the former employee's access to the trade secret information, and the trade secret intellectual property rights in the information will be lost.

Real-World Scenarios I: Inadvertent and Careless Disclosure

We have talked about how the four required proofs can be undermined in somewhat hypothetical terms. How do these considerations translate into real-world situations? The following scenarios are common ways in which trade secret property rights are lost through carelessness or failure to be a good steward of trade secret information. We will discuss scenarios of inadvertent and careless disclosure first.

Inadvertent or careless disclosure of the company's trade secret information destroys the trade secret property rights of the information. How can the information qualify as a trade secret if the information owner or his agents — his employees — disclose the information outside the bounds of a confidential relationship? Such disclosure takes the information out from under the protection of the UTSA in two ways. Careless or inadvertent disclosure makes the information generally known, disqualifying it for trade secret protection under the UTSA definition. Careless and inadvertent disclosure also proves that measures appropriate under the circumstances to maintain the secrecy of the information have not, by definition, been taken: the most basic security measure of all is the obligation of the trade secret owner not to knowingly disclose the information outside of a confidential relationship.

Common scenarios for careless and inadvertent disclosure

include the following.

The Trade Show

Many business sectors have annual trade shows in which companies display their products or services to actual and potential customers. At these trade shows they often also display their trade secrets to actual and potential competitors. Have the company's booth personnel, often marketing and engineering personnel with detailed knowledge of the company's trade secrets, been instructed to limit their discussions with booth visitors to public information? Or are they likely to disclose company trade secrets in order to impress a customer with the company's commitment to enhancing its products and services?

The Conference Speech

Industry conferences and seminars feature speakers that inform attendees about new developments in the field. The company's senior research, development, and engineering employees are prime candidates for speaker status at these events. Have the company's personnel speaking at these events been instructed as to what information can and cannot be disclosed in this public venue? Have their presentations been vetted for non-disclosure of the company's trade secret information by the staff assigned to the protection of the company's trade secrets? Or are they likely to disclose too much of what they are working on in order to impress their peers at these events?

The Sales Call

The company’s sales force meets with customers every day, and their income is often dependent on their sales success due to commission-based compensation packages. The company's sales force is also often aware of trade secret information being developed and used within the company. Has the company's sales force been instructed as to what information can and cannot be released to potential customers in closing the sale? If sensitive information is authorized to be released, does this disclosure to the potential customer occur after a non-disclosure agreement has been executed? Or is the sales force likely to divulge proprietary information to a potential customer in order to close the deal and capture the customer's business?

A variation of the sales call scenario is the sales call in which company management, marketing, or engineering personnel accompany the sales force in making the call on the customer. Company staff likes to get out of the office, take the unaccustomed business trip, and meet with real or potential users of the company's products and services. Have these personnel been instructed to limit their presentations and discussions to publicly disclosed information? Have their presentations been vetted for non-disclosure of the company's trade secret information by the staff assigned to the protection of the company's trade secrets? Or are they likely to discuss sensitive information, with which they have daily contact at the home office and in the development of which they are justifiably proud of their own contributions, in an effort to be helpful to the sales force in impressing the customer?

The Customer Visit

This is the converse of the sales call — the actual or potential customer visits the company. The risks of disclosure of trade secret information during the customer visit are similar in nature to the sales call but greater in extent due to the increased access to the company's staff and premises. In addition to the considerations already discussed for the sales call, there are several more that relate to the customer visit. Has the customer's visit been carefully orchestrated? Have the areas in the company to which the customers are to be admitted been carefully screened? Is trade secret information absent from or concealed in those areas? Has the staff assigned to the protection of the company's trade secrets been involved in the planning and implementation of customer visits?

The Users Group Meeting

Users group meetings are customer visits multiplied by the number of customers represented, and the risks of disclosure are multiplied as well. All of the cautions already discussed for sales calls and customer visits apply, but also, due to the increased risk, users group meetings should be held off-site. It is too difficult to control dozens or hundreds of simultaneous visitors to the company's premises. Instead, offer carefully chaperoned tours of the facilities to smaller groups of attendees at specific times and in non-sensitive areas only. Do not forget that in many customer purchasing decisions, there are customer staffers who argued in favor of making the purchase from other vendors. They are now members of your users group, but may also remain active partisans of your competitors.

The Employment Interview

The employment interview might often be called "The Competitor Visit". Often the company's best prospects for new employees currently work for the company's direct competitors. Their general knowledge, skills, and experience, which they are free to use in their further employment, directly relate to the business and technical areas in which the company and its competitors operate. However, interviewers must not forget that, after the interview, the candidate will go back to his current job with the company's competitor. There are no guarantees that the individual has a real interest and will actually accept an offer from the company; he may simply be fishing for information. Are the company's interviewers aware of the dangers of unprotected disclosure of the company's sensitive information during the employment interview process? Will they limit their discussion of the position's job responsibilities to general terms? Or are they likely to discuss sensitive specifics of the program for which the candidate is interviewing in an effort to excite him about the job and convince him to sign on as an employee?

The Media Interview

Perhaps no inadvertent or careless disclosure is quite as egregious as the disclosure during a media interview, in which the sensitive information is disclosed to someone whose stated purpose is to write up the information, print it, and distribute it to the world. The best advance solution to this potential problem is to grant the interview under the aegis of a non-disclosure agreement, with a right of review of the pre-publication text for the specific purpose of ensuring no sensitive information is included. If this is not possible, natural questions arise. Are the company interviewees aware of the dangers of unprotected disclosure of the company's sensitive information during the media interview process? Will they limit their discussion of the company's business to public information? Or are they likely to succumb to the reporter's encouragements to provide them with a 'scoop'?

Spies

This section would not be complete without a discussion of spies — competitive intelligence agents. As we mentioned in Chapter 2, the Annual Report to Congress on Foreign Economic Collection and Industrial Espionage—2004 noted, "Most of the foreign entities attempting to acquire US technology last year employed tools and techniques that were easy to use, inexpensive, low risk, and sometimes legal. In a majority of cases, foreign collectors simply asked—via e-mail, phone call, FAX, letter, or in person—for the information." Competitive intelligence agents — whether employed by foreign entities or by the company's domestic competitors — can legally collect sensitive information that is carelessly or inadvertently disclosed by the company's employees using any of the scenarios outlined in this section. Posing as customers at trade shows, attending conferences, interviewing for employment at the company, and posing as reporters conducting interviews for trade publications are common and legal activities targeted to eliciting the company's sensitive information.

Real-World Scenarios II: Unprotected Disclosure

For purposes of discussion, we will distinguish unprotected disclosure from careless or inadvertent disclosure by intent. In unprotected disclosure, the disclosing party intends to disclose the information to the other party, but does so without first establishing a burden of confidentiality on the other party. A burden of confidentiality with the other party can be created by hiring the individual as an employee, or by the execution of a non-disclosure agreement between the company and the other party. Unprotected disclosures usually take place when the disclosing party intends to create such a confidential relationship in the future, but has not yet done so, or when the disclosing party does not realize that such a step is necessary. Like careless and inadvertent disclosures, unprotected disclosures destroy the company's trade secret property rights in the disclosed information, even if a confidential relationship covering further disclosures is established later.

One twist on confidential relationships sometimes occurs in real situations. The courts may determine from the specifics of the case at hand that a confidential relationship was established by the circumstances of the disclosure. That is, even without a non-disclosure agreement, the court may find that the parties had agreed to treat the information disclosure as confidential, or that a reasonable man would have considered the relationship confidential given the totality of the circumstances. This scenario came about in *Learning Curve Toys, Inc. v. Playwood Toys, Inc.*, an interesting decision that we discussed earlier and which is contained in Appendix B.4. Nevertheless, it is much better to rely on a written instrument establishing a confidential relationship than on the vague and sometimes conflicting recollections of witnesses to a verbal agreement.

Common scenarios for unprotected disclosure include the following.

<u>Contract "Employees"</u>

It should be noted that hiring an employee creates a confidential relationship only if the employee is a true, W-2 employee. That is, a statutory employee under labor laws, and for whom the company provides employee benefits and performs income tax, Social Security and Medicare withholding. So-called contract "employees" are not employees in the legal sense, and there is no confidential relationship created when a contract employee is hired.

The distinction is important. Disclosure of the company's trade secret information to a contract employee prior to or absent the execution of a non-disclosure agreement constitutes an unprotected disclosure and any trade secret property rights in the disclosed information are lost. Contract employees should not be allowed to begin working within the company until the non-disclosure agreement is executed. If the contract employee wants his attorney to look at the agreement first, that's fine, but he cannot begin working on contract for the company until after his attorney's approval is obtained and the agreement executed. Until that time, the contract employee should be treated as a visitor, escorted at all times when on company premises, and no trade secret information should be disclosed. No project is in such immediate need of extra manpower as to allow the company's trade secret information to revert to the public domain, free for use by anyone.

Preliminary Discussions

A potential customer will sometimes wish to hold "preliminary discussions" for which the "formality" of a non-disclosure agreement "isn't really necessary" or "is premature". The customer may be interested in the company's products or services for internal use, for incorporation in the customer's own product or service offerings, or for resale. What the desire to avoid a non-disclosure agreement often signals, however, is that the potential customer has not yet decided whether to purchase the product or service under discussion from outside or to develop it themselves. They do not want to be restricted from the independent development of similar information if they decide to undertake the development themselves, and besides, the company's representatives may tell them more than they should.

Under no circumstances should trade secret information be disclosed before a confidential relationship is established. The company's representatives must not disclose any non-public information to the potential customer in an effort to win the sale. They are educating a potential competitor and destroying trade secret property rights in the disclosed information. Even if the potential customer decides ultimately to buy the product or service, they may buy it from one of the company's competitors, with whom they may share the forfeited trade secret information. The company's representatives should instead make a compelling presentation of the company's publicly available information about the product or service, and push for a non-disclosure agreement to allow discussions to proceed to proprietary information.

Employee Interviews and Resumes

Employees have the right to use their general knowledge, skills, and experience in pursuing other employment, but they do not have the right to use or disclose their employer's trade secret information. The law prohibits the disclosure of the trade secret information of current or past employers in resumes, on job applications, and during interviews. Nevertheless, resumes are often filled with detailed descriptions of projects and technologies in which the employee was active, and these resumes are submitted to Monster.com and other job sites for the whole world to see. Further disclosures of trade secret information often occur during interviews, with the interviewee seeking to impress his interviewers by describing in detail the projects and technologies he worked on.

Note that these are not careless and inadvertent disclosures. The employee is using the company's trade secret information for personal economic gain in the form of a new job. His ties of loyalty to his current employer are already weak, or he would not be out pursuing new employment. He has decided that his personal gain in a new job outweighs the company's intellectual property rights in its trade secrets, and intentionally discloses those trade secrets to impress potential employers.

The employee here is legally liable for the unprotected disclosure of the company's trade secrets, which constitutes a breach of his fiduciary obligation to keep his employer's proprietary information secret. It is important to note, however, that the person to whom the employee disclosed this information has not done anything wrong. The employee destroyed the company's trade secret property rights in the information. Inducement to breach a fiduciary obligation is actionable, but unless the potential new employer offered some incentive beyond the customary recruiting process, it is the employee who is solely at fault.

This unprotected disclosure is particularly egregious because the company's competitors will often be prime employer candidates in an employee's job search. The employee's general skills, knowledge, and experience are directly applicable to the competitor's business needs as well. Competitive intelligence agents take advantage of employee's willingness to disclose proprietary information during job searches by surfing the net for employee resumes, and also often pose as headhunters conducting telephone interviews of company employees.

The Simultaneous Disclosure

In the press of business, companies are tempted to take short cuts. One short cut that can destroy trade secret rights with an unprotected disclosure is the simultaneous release of proprietary information with the transmittal of the non-disclosure agreement. The typical case occurs when the parties have agreed to the execution of an NDA and the transmission of the information. The disclosing party, trying to save time and get the information-dependent activity underway, transmits the NDA to be executed and the proprietary information as attachments in the same email. The transmission of the information in this context is an unprotected disclosure because the NDA has not yet been executed. The courts may determine that a confidential relationship already exists because the execution of an NDA was agreed upon prior to the disclosure. However, the courts may decide that the confidential relationship does not yet exist, or does exist but in more general terms, because the written NDA was not yet executed and the specific terms of the NDA were not discussed prior to its execution. Information disclosed simultaneously with the transmission of the NDA may then be judged an unprotected disclosure, and trade secret property rights in the information forfeited.

Unprotected Disclosure Within an NDA

Finally, it is all too common for companies to make unprotected disclosures within a non-disclosure agreement. NDAs usually specify the conditions that apply to a disclosure of information in order for the information to be protected as a trade secret under the NDA. Common terms of an NDA include that the disclosing party must notify the recipient as to which specific information is proprietary, that any written disclosure of proprietary information must be marked Proprietary and Confidential, and that any verbal disclosure of proprietary information be followed up with a written and marked copy within one week. Any other disclosure of information falls outside of the scope of the NDA and is an unprotected disclosure resulting in forfeiture of trade secret property rights.

A company's concern is usually whether there is an NDA in place, but little attention is paid to its terms in practice. Failure to comply with the terms of the NDA, however, removes the disclosure from its protections and destroys trade secret intellectual property rights in the information.

Real-World Scenarios III: Failure to Take Reasonable Measures

Failure to take reasonable measures can result in the loss of trade secrets in two ways. First, the trade secret information may actually be accessed through the lack of reasonable measures, such as in the case of a dumpster-diving competitive intelligence agent obtaining company trade secret information from publicly available trash bins. This is the scenario of which most companies are aware and which they take the most steps to counteract.

In the second case, the trade secret information may have been accessed by other means, such as by an employee to whom the information was disclosed in order to enable his performance of his job duties, and who then argues at trial that failure to take reasonable measures by the company has resulted in a forfeiture of trade secret rights by the company. In this case the failure to take reasonable measures was not the immediate cause of the information loss, but may ultimately cause the loss of trade secret property rights in litigation.

There is no limit to what a defendant in a trade secret misappropriation case will contend amounts to a failure to take reasonable measures. If the company password protects all of its computers, defendant will claim that a firewall should also have been part of reasonable measures. If the company uses passwords and a firewall, defendant will claim that dongles should also have been part of reasonable measures. If the company uses passwords, firewalls and dongles, defendant will claim that biometric devices should also have been part of reasonable measures. Defendant will make these claims even though there was in fact no access to the trade secret information through any lack of security measures: the information was disclosed to defendant in order to enable his performance of his job duties.

On the other hand, there is no limit to what plaintiff, caught unprepared by the need for litigation, will desperately claim to be adequate security measures. The mere presence of locks on doors, trade secret information kept in a jumble of papers in the company owner's desk rather than organized and filed, the obscure physical location of the company, even the lack of a trade secret inventory, thereby providing no trade secret list to be stolen, have been claimed as "reasonable measures".

The courts will brush aside the extreme claims from both sides and focus on what was "reasonable under the circumstances". What is the standard of care for trade secret information in the industry? Have there been information losses from the company before? Were security procedures tightened in response? How big is the company? Is it large enough to have "large corporation" measures in place? Or is it a small mom-and-pop, for which such measures would be a financial burden?

The company should have two goals in designing trade secret security measures. The first is to prevent information loss through security lapses, and most companies already design their security measures toward this goal. The second is to ensure a favorable result in trade secret litigation, including a court finding that reasonable measures were taken. This is where companies typically fail.

The next several chapters will discuss security measures in great detail. Their importance in maintaining the company's trade secret property rights in its information cannot be overemphasized.

Summation

Unpatented business information that is not protected as a trade secret is free and open for use by anyone for any legal purpose. There are many ways to forfeit trade secret property rights in information, but the most common is a failure to prove the existence of a trade secret through failure to take reasonable measures to protect the secrecy of the information. Inadvertent and careless disclosures destroy trade secret property rights in information, as does unprotected disclosure to third parties, as these disclosures are by definition failures to take reasonable measures to maintain secrecy. Finally, the failure to take reasonable measures to protect the secrecy of the information can result in the loss of trade secret property rights even if the actual access to the information was proper and not due to any lapse in security measures.

PART II: SECURITY

The future of the nation depends in no small part on the efficiency of industry, and the efficiency of industry depends in no small part on the protection of intellectual property.

— United States Court Of Appeals For The Seventh Circuit

Chapter 5
Introduction to Security

Executive Summary

Information security is a dynamic environment of measures and countermeasures in the effort to protect the company's trade secrets. Considering a model of the company as a fenced enclosure can aid in the discussion of information security. In formulating security measures, it is important to distinguish between insiders and outsiders, and between appropriation and misappropriation. Security of trade secrets in the international marketplace is evolving rapidly and will require interim solutions that adapt to new conditions.

The Dynamic Security Environment

While the law of trade secrets discussed in the previous chapters is well-established, the areas of information threat and information security are evolving rapidly. New hardware, software, and business methods for protecting information are being developed and released almost daily to counter the growing threat of information theft. As the methods for stealing trade secrets are continuously improved, the methods for preventing the theft of information are evolving to meet these threats in a cat-and-mouse game of prevention and circumvention. Consequently, were we to discuss specific methods of information security in these chapters, the recommendations would be obsolete before this book is even published.

Instead we will concentrate on the general methods of trade secret access and theft within which the specific threats fall. Just as the chapters on the law of trade secrets were not intended to educate the reader in the law so much as to provide the background required for effective communications with attorneys, the chapters on the security of trade secrets are not intended to educate the reader in information security so much as to provide the background for effective communications with information security specialists.

The Fenced Enclosure Model

It can be helpful in discussing trade secret security issues to consider the company as a fenced enclosure. The fence represents the company's security measures, and in general encompasses the company's premises. Proprietary information is developed, used and stored within the fenced enclosure. Insiders move freely in and out of the enclosure, while outsider entry and exit from the enclosure is carefully controlled.

Given this model, it is possible to see the mechanisms by which access to trade secrets can be accomplished. Certainly insiders have access to the enclosure and the proprietary information within it. Insiders move in and out of the company freely, and proprietary information moves with them, in their minds, in portable computers, and in media such as drawings, CD-ROMs, and USB flash drives. The company has a network connection that penetrates the fence, and through which proprietary information can be transmitted, as well as outbound mail, courier and parcel shipping. Finally, outsiders are allowed into the enclosure, usually with badge systems and escorts, but they may gain access to proprietary information while inside the company if the visit is not properly orchestrated.

Certainly it would be easy to keep the company's proprietary information secure if all transmission of information through the fence and out of the enclosure were eliminated, but this is impractical. Cutting the Internet connection, eliminating outbound mail, courier and parcel services, eliminating outsider visits to the company, and prohibiting the transport of proprietary information outside of the company would eliminate security problems, but would also destroy the ability of the company to do business as well as the economic advantage derived from trade secret information. In this scenario, proprietary information would be a burden that would cripple the company's ability to do business while conferring no advantage at all. Besides, the company still has to let the employees go home in the evening, and they carry the company's proprietary information in their heads.

What the company can do, however, is to strictly limit the transmission of proprietary information through the fence and out of the enclosure to those activities necessary to the business goals of the company.

Distinguishing Between Insiders and Outsiders

In any discussion of trade secret security, it is important to distinguish between insiders and outsiders, between the internal and external threats to trade secrets. Insiders are individuals and organizations that have a legal obligation of confidentiality to the company. Insiders include employees, who have a fiduciary obligation under agency and employment law to hold the company's proprietary information in confidence and not to copy, disclose, or use the information for their own benefit or the benefit of others. Insiders also include third parties, including contract employees, consultants, suppliers, and customers, who have a contractual obligation to hold the company's proprietary information in confidence in accordance with the terms of a non-disclosure agreement executed between the contractual third party and the company. Finally, insiders include third parties who have express or implied fiduciary obligations to hold the company's proprietary information in confidence due to ethical or legal obligations deriving from their relationship to the company, including banks, attorneys, employee healthcare providers, and security firms such as information disposal companies.

Outsiders, then, encompass literally everyone else, those individuals and organizations who are strangers and who have neither a fiduciary nor a contractual obligation to hold the company's proprietary information in confidence. Outsiders include the general public and more sophisticated parties including competitive intelligence professionals, hackers, media, and competitors, as well as contractors, consultants, suppliers, and customers not bound by a non-disclosure agreement.

Distinguishing Between Appropriation and Misappropriation

It is important in discussing information security to distinguish between appropriation — the legal use of information — and misappropriation — the illegal use of information that provides the company with the recourse of a legal cause of action. Misappropriation often depends on whether or not the trade secret was acquired by proper means.

The Uniform Trade Secrets Act (UTSA) includes a definition of improper means, shown in Table 5-1.

Table 5-1 UTSA Definition of Improper Means

> "Improper means" includes theft, bribery, misrepresentation, breach or inducement of a breach of a duty to maintain secrecy, or espionage through electronic or other means.

Note that all of the improper means listed are in themselves illegal. These provide a basis for criminal prosecution for the predicate offense in addition to a cause of action for trade secret misappropriation. The UTSA also provides a separate definition of misappropriation, shown in Table 5-2.

Note that paragraphs (i) and (ii)(B)(I) do not excuse acquisition, disclosure, or use of a trade secret acquired through the use of improper means by another if the receiver knows or has reason to know that improper means were employed in obtaining the information. Paragraph (ii)(A) forbids disclosure or use by a person employing improper means to obtain the trade secret, such as a thief or hacker. Paragraph (ii)(B)(II) notes that the circumstances of acquisition can give rise to an obligation of confidentiality. Paragraph (ii)(B)(III) addresses misappropriation by a person who has an obligation of confidentiality, the insiders to whom proprietary information is disclosed under an NDA or fiduciary obligation. Finally, paragraph (ii)(C) notes that, in the case of a disclosure by accident or mistake, if the person receiving the information becomes aware that the information is a trade secret before a material change in his position, then disclosure or use constitutes misappropriation. Any copying, disclosure, or use of information outside of this definition is not a misappropriation and is not actionable by the company under the UTSA.

Most information security measures target improper means to prevent company proprietary information from being lost. This is necessary but not adequate. We will discuss more complete security measures in the next few chapters.

Table 5-2 UTSA Definition of Misappropriation

"Misappropriation" means:

(i) acquisition of a trade secret of another by a person who knows or has reason to know that the trade secret was acquired by improper means; or

(ii) disclosure or use of a trade secret of another without express or implied consent by a person who

(A) used improper means to acquire knowledge of the trade secret; or

> (B) at the time of disclosure or use, knew or had reason to know that his knowledge of the trade secret was
>
> > (I) derived from or through a person who had utilized improper means to acquire it;
> >
> > (II) acquired under circumstances giving rise to a duty to maintain its secrecy or limit its use; or
> >
> > (III) derived from or through a person who owed a duty to the person seeking relief to maintain its secrecy or limit its use; or

> (C) before a material change of his [or her] position, knew or had reason to know that it was a trade secret and that knowledge of it had been acquired by accident or mistake.

International Security of Trade Secrets

The chapters on security are written toward security of trade secrets in the domestic United States. Other countries have their own trade secret laws, which may be similar to those of the United States or very different. Some of these laws are strictly enforced, such as in Japan, while others are loosely enforced if at all, such as in China. As measures for security are adapted to the legal milieu in which they are applied, international security for trade secrets requires careful consideration by professionals with experience in these markets.

The security of trade secrets in China is currently of particular concern to U.S. executives. The situation is changing rapidly here, as China strives to address the intellectual property concerns being advanced by the developed nations. The Chinese government is tightening intellectual property protections within China, but the initial impetus is limited primarily to patents and copyrights. Trade secrets protection in China is lagging, and information owners must be careful. Establishing a subsidiary in China is currently popular, but this decision must be weighed against the serious risk of loss of proprietary information with no right of immediate judicial redress.

The focus here is on domestic security for trade secrets. Many of these methods and cautions will be equally applicable overseas while some will not be. Any company with international operations must consult with professionals in the respective countries to tailor the most effective information security and trade secret protection program on a country-by-country basis.

Summation

The dynamic information security environment is adapting constantly to meet new information security threats. In assessing these threats and the appropriate countermeasures, it can help to consider a model of the company as a fenced enclosure. It is also important to distinguish between insider and outsider threats to information. The UTSA defines misappropriation of a trade secret; all other uses of information are proper under the UTSA. International security concerns will require professional guidance with specific experience in the particular countries of concern to fashion a program that can adapt to changing international conditions.

Chapter 6
Security Against Outsiders

Executive Summary

Outsiders attempt to access company proprietary information through both proper and improper means. Outsider access to trade secrets by proper means is the result of carelessness on the part of company employees. The only remedy to outsider access to trade secrets by proper means is employee education. Outsider access to trade secrets by proper means results in the loss of the company's trade secret property rights in the information. In contrast, outsider access to trade secrets by improper means involves an illegal or improper act such as fraud or theft. Outsider access to trade secrets by improper means provides the remedies of criminal prosecution for the illegal act as well as a civil lawsuit for trade secret misappropriation. Outsider access to trade secrets through hacking is particularly dangerous because of the potential for loss of the company's entire trade secret portfolio within minutes. Red team attacks can test the company's security measures against outsiders and provide valuable training for employees.

Outside Access by Proper Means

The only way for an outsider to access the company's proprietary information by proper means is for the information to be made available without restrictions to the outsider by the company or someone working for the company. These leaks take many forms but most often fall into the categories listed in Table 5-1.

Careless, Inadvertent and Unprotected Disclosures

The nature of careless, inadvertent, and unprotected disclosures was discussed at some length in the Chapter 4. The only security measure that is effective against careless and inadvertent disclosures is employee education. Employees need to know what constitutes careless, inadvertent, and unprotected disclosure, and what the company's trade secrets are, in order to prevent such disclosures. This cannot be a haphazard or partial education of some employees, for the security of the company's proprietary information depends on the weakest link. Every employee who has knowledge of any of the company's trade secrets needs careful training and reinforcement on avoiding careless, inadvertent, and unprotected disclosures.

Table 5-1 Typical Ways Proprietary Information Is Made Available For Access By Outsiders

- Careless or inadvertent disclosure during trade shows, conferences, sales calls, interviews
- Unprotected disclosures to potential customers, contract employees, potential employers
- Discussion of proprietary information between insiders in public locations
- Errors in the transmission of proprietary information in the mails and over the Internet
- Careless disposal of company records, documents, computers, and storage media

It is also important to mention here that misrepresentation by a competitive intelligence agent of his identity or motives, absent more, does not constitute an illegal act. At a trade show, whether a competitive intelligence agent represents that he is a professor for "Central States University" doing research, or a security professional for "ABC Corporation" gathering information for an internal security effort, or a disaffected employee for "Competitor X" seeking a new job makes no difference. If the information is a trade secret, the company insider should not make an unprotected disclosure of the company's proprietary information to any third-party recipient regardless of his or her true identity and motives.

Whether the agent misrepresents his identity and motives or comes right out and says he has been retained by the company's competitor to obtain proprietary information by all legal means is not material: the disclosure should not be made in any case. Employees will certainly recognize the latter scenario as being dangerous, but they should be educated in the proper handling of proprietary information to recognize the former situation as being dangerous as well.

Discussion of Proprietary Information in Public Places

Discussion of proprietary information between insiders in public places is a serious problem. By public places we mean any situation in which outsiders are present, including meetings in hotel lobbies, airport terminals, and restaurants, social gatherings on public or private premises where insiders may "talk shop", and travel on public carriers. There is no obligation on the part of outsiders who may overhear the conversation to block their ears or step out of earshot, nor is there any prohibition on them copying, disclosing, or using the information for their own benefit or the benefit of others.

The same cautions apply to cell phone conversations conducted in public places. In this case, only one side of the conversation is disclosed to outsiders who may be present, but this again constitutes a public disclosure. This particular case is aggravated by people's tendency to speak overly loudly into cell phones, and to be oblivious to the environment around them. As a consequence, they may be shouting the company's trade secrets across the airport terminal for all to hear.

A specific case will illustrate the danger. When traveling to the industry's biggest annual trade show twenty years ago, one of the authors was seated quite coincidentally behind two sales representatives of a competing firm on their way to the same trade show. During the three-hour flight, these two sales representatives discussed in detail their company's strategy for the show, including their strategy for selling against the author's company. One even mentioned to the other not to offer or attempt to demonstrate a much-anticipated new product they were introducing at the show, because it did not really work yet. Armed with this information about this "vaporware," the author's company thwarted the competitor's sales activities at the show and the competitor never knew how it happened.

The competitor's employees discussed extremely sensitive proprietary information in a public place where outsiders were present. They acted no differently than they might if they were in their offices. Speaking in normal voices, they informed everyone within earshot on the airplane, including a competitor, of their "secret" plans for the show. Any proprietary rights in this information were forfeited. The recipients within earshot had no obligation of confidentiality. In fact, your author's fiduciary duties to his own employer required him to immediately disclose this "competitive intelligence" information to his employer at the trade show. Would these sales people have held such a public discussion of sensitive information if they had known that a regional manager for their biggest competitor was seated behind them on the plane? Of course not! But they did not know that the competitor was not there, and that's the point. Proprietary information should not be discussed in public places. Such discussions create a very serious risk that trade secret rights will be forfeited without any knowledge whatsoever to the trade secret owner, and without any recourse whatsoever.

<u>Errors in Transmission</u>

Occasional errors in transmission are inevitable in any medium. The misaddressed email is particularly common, where an employee sending an email containing proprietary information enters, for example, the email address FredSmith412@aol.com instead of FredSmith421@aol.com. Hopefully, "FredSmith412" doesn't work for a competing firm, and will simply discard the erroneous transmission. Because the erroneous addressee is unlikely to have an interest in the information, this source of loss of proprietary information is less likely to have harmful effects, but the possibility is there.

Mistakes happen, and it is impossible to protect against all errors in transmission by simply exhorting people to be careful. Transmissions of proprietary information out of the firm should be strictly limited to those necessary to further the company's goals to limit the exposure. In addition, transmissions of proprietary information should be encrypted to ensure that only the correct recipient can access the information. Systems for digital rights management (DRM) are evolving rapidly to make this process transparent to users while prohibiting transmission of proprietary information "in the clear".

In all transmissions of proprietary information, encrypted or otherwise, the information should be prominently marked "Proprietary and Confidential" so that, per the UTSA definition of misappropriation, the recipient of information due to an error in transmission "knew or had reason to know that it was a trade secret and that knowledge of it had been acquired by accident or mistake."

Careless Disposal

The final way proprietary information is made available for outsider access by proper means is the careless disposal of company records, documents and storage media, which results in serious losses to many companies. The company needs to properly dispose of all items containing proprietary information, including research prototypes and manufacturing equipment, to ensure that they are not accessible to outsiders once they leave the company. The company's waste stream is a primary target for competitive intelligence agents seeking the company's proprietary information. Obtaining such information from the company's discards after they have become publicly accessible is not a crime, no cause of action exists, and trade secret intellectual property rights in the information are lost. "Wastepaper archeology" is a favorite technique among competitive intelligence professionals.

In securing against this loss, employee education is not enough. The company must provide the necessary means for proper disposal of items containing proprietary information. Locked document disposal bins should be located throughout the company, and emptied regularly by a bonded outside firm that performs on-site shredding. Similar bins can be provided in the same location for the destruction of CD-ROMs, flash drives, floppy disks, and hard drives containing proprietary information. Many on-site shredding firms are able to accommodate these devices in the same bins as paper documents. An internal phone number should also be posted in these same locations for the secure pickup and disposal of larger items such as research prototypes and discarded manufacturing equipment. Failure to provide secure means for employees to destroy items containing proprietary information constitutes failure to take reasonable measures under the circumstances and results in the forfeiture of trade secret property rights in the information.

A special note about magnetic media is appropriate here. Deletion of data on hard disks does not actually result in the erasure of the data. Instead, the data is marked deleted but remains on the hard drive in unallocated sectors. The "erased" data can be retrieved with specialized software. Similarly, reformatting of hard disk drives does not result in erasure of the information that the drive contained before the reformatting occurred. Specialized software and hardware can discern between "0 now and was a 0 before" and "0 now but was a 1 before", and between "1 now and was a 1 before" and "1 now but was a 0 before". Even a single overwrite of the hard drive thus leaves the overwritten data recoverable by outsiders who may gain access to the drive after it leaves the company's premises. Secure erasure programs can perform multiple overwrites to hard drives to ensure that data is unrecoverable. However, the best solution is simply to destroy the discarded hard drives and other electronic media containing any fragments of proprietary information. While some companies donate their used computer equipment to charities, this is not recommended for the reasons discussed above. The company should destroy its old computers and give the charity a cash donation instead. It is more useful to the charity, and can be much cheaper for the company in the long run.

Outsider Access by Improper Means

Illegal access to proprietary information is what people usually think of when they hear about trade secret theft, and consequently it is the area where companies have historically focused their efforts to protect intellectual property. Illegal access can occur through fraud, trespass, theft, hacking, and inducement of an insider to breach their fiduciary or contractual obligation of confidentiality. The company's recourse on discovery of illegal access to proprietary information includes both criminal prosecution for the illegal activity and a civil trade secret misappropriation lawsuit.

These recourses, however, are only available if the company is aware of the theft, and if the company can identify the perpetrators. Trade secret theft is unique in that the theft can occur without anything missing and without the company being aware of the theft. Theft of a physical asset such as a piece of machinery involves its removal from the company and its absence will be noticed. Theft of intellectual property can involve copying of the information, in which case there is nothing missing to indicate the theft occurred. Since a trade secret derives its value from being secret, a theft that destroys the secrecy of the information results in the loss of its value to the company even though the information is still in the company's possession.

Securing trade secrets against illegal access by outsiders necessarily includes three parts. First, the company should take measures to make the illegal access difficult. Second, the company should take measures to ensure that the illegal access to the trade secret is noticed. Finally, the company should take measures to ensure that the illegal access to the trade secret creates evidence that allows the company to identify the perpetrator.

Access By Fraud

Access by fraudulent means occurs when an outsider falsely represents himself to be an insider for purposes of getting another insider to disclose proprietary information. The fraudulent representation is an improper means, and the resulting access to proprietary information is an actionable misappropriation.

An example of access by fraud would be an outsider presenting a forged business card and claiming to be an employee of a company in a confidential relationship with the trade secret owner. The impostor may even represent himself to be an employee of another division of the trade secret owner or a member of the sales staff in a remote office. Relying on the fraudulent misrepresentation to induce disclosure of trade secret information, the trade secret owner is duped. This fraudulent misrepresentation is actionable.

Contrast this with the trade show situation discussed in a previous chapter, where there was no misrepresentation of a confidential relationship. In this case, the misrepresentation relates solely to the identity of a third-party outside a confidential relationship. In these circumstances, there is no actionable fraud. The competitive intelligence agent who claims to be associated with a company or organization that does not have an obligation of confidentiality with the disclosing company has in fact accurately identified himself as an outsider, to whom proprietary information should not be disclosed. In the case of access by fraud, however, the outsider has fraudulently represented himself to be an insider in order to induce a disclosure that would not be appropriate given his actual relationship to the company.

Prevention of access by fraud focuses on making sure that any person to whom trade secret information is disclosed has an obligation of confidentiality to the company. Strong efforts to ensure the person is who he claims to be and execution of an additional non-disclosure agreement on the spot are methods of achieving this requirement. Better is to not disclose proprietary information in any ad hoc situation, but to make such disclosures in better planned and managed activities such as invitation-only presentations at trade shows and other industry events.

Access by Trespass

Access by trespass occurs when an outsider makes unauthorized entry to the company's property to gain access to proprietary information. Unauthorized access can include trespassing on the company's property to take pictures of items that are not visible from other properties, wandering into unauthorized areas during a plant trip or company visit, and, in extreme cases, breaking and entering the company's premises. The common movie scenario of the agent breaking into a building to photograph secret documents with a tiny camera constitutes access by trespass. The trespass is improper means, and the resulting acquisition of proprietary information is an actionable misappropriation.

Prevention of access by trespass is best accomplished by making unauthorized access to the company's premises difficult. Access controls, alarms, motion detectors, and cameras limit access and prevent unauthorized entry. It is also important to properly chaperone outsiders on company premises so that trespass of forbidden areas does not occur during authorized plant visits.

Detection of access by trespassing is performed through monitoring of the company's premises. Security cameras should continuously record activity in sensitive areas, and security tapes or video files should be archived. The financial and operational burden of this measure can be reduced by limiting the areas in which highly sensitive information is developed, used, and stored. Motion detectors in vacant areas, such as the shop floor after hours, can be used, in combination with security guards, to detect trespassers with immediate apprehension of the violators.

Video security tapes or files will insure the identification, apprehension, and conviction of any trespassers. Knowing that a trespass occurred and that proprietary information may have been accessed does little good without the means to identify the trespasser. In addition to the archiving of security tapes and files, other types of access records such as visitor logs and ID reader logs should be retained by the company. Since trade secrets may exist indefinitely, logs that record access to locations in which trade secrets are developed, used, and stored should likewise be retained indefinitely.

It is important in all cases of access by trespass to report the trespass immediately to authorities. There must be a zero tolerance policy. All trespassers must be prosecuted. Companies are often reluctant to report such incidents to the police. Realize that the trespass put all of the company's proprietary information at risk. The police have resources for the apprehension and prosecution of trespassers that are not available to the company that proceeds with just a civil lawsuit. In addition, if there is a criminal conviction, then liability will be established as a matter of law in the subsequent civil action, and the only issue will be the amount of damages and the scope of injunctive relief. Even without a criminal conviction, juries expect trespass incidents to be reported to police, and the failure to report the crime may impact the civil action if this fact becomes known to the jury.

Access by Theft

Access by theft occurs when an outsider steals an item or items containing proprietary information. The theft can occur inside the company's premises during a company visit, such as when a visitor surreptitiously removes documents or media off an unoccupied desk during a plant tour, or the theft can occur off-site, such as when a laptop is stolen from an employee during a business trip In either case, the theft is illegal, and the resulting access to proprietary information is an actionable misappropriation.

Preventing access by theft has two components depending on whether the theft is on-site or off-site. To protect against access by theft on-site, outsiders should be chaperoned by an employee at all times when on the company's premises. They are inside the fenced enclosure where trade secrets are being developed, used, and stored, and visitors must not be allowed to wander freely.

Protecting against access by theft off-site is more difficult. Employees carry proprietary information in their computers, inside their briefcases, and in their cars, and all three are targets for thieves. Usually these thieves are only interested in the physical item and not in the information it contains, but the risk of disclosure is still very high. Who's to say what a thief will do when he realizes the information contained in the briefcase is more valuable than the briefcase itself? Even more worrisome, there have been some cases of targeted theft of these items by agents seeking the proprietary information of the company.

The theft of laptop computers has become an especially troublesome source of loss of proprietary information. Several high-profile cases appeared in the newspapers during the writing of this book, and the costs of both damage control efforts and reputation for the companies involved has been high. Employees must be cautioned that their cars, briefcases, and especially laptop computers are targets for thieves, and that the proprietary information they contain is at risk of disclosure from such a theft. Executives need to get out of the habit of routinely carrying around large amounts of highly confidential company information in their laptops and briefcases.

There are several methods for preventing and detecting access by theft or limiting the impact if such theft and access occur. One method is limiting the use of paper documents and encrypting sensitive files contained on laptop computers to prevent access if a theft occurs. Additionally, methods can be used to track stolen items and recover them before access is achieved, such as by using GPS tracking equipment on executive briefcases and laptop computers. One method for limiting the impact of a theft is limiting the amount of information that employees carry out of the company, which can be done using checkout procedures, using corporate file servers to limit the information on individual laptops, and even by such simple measures as splitting up large documents into smaller pieces which can be worked on and used individually. The technology for preventing, detecting, and reducing the impact of access by theft is developing rapidly, and current methods should be researched carefully before selecting a suite of such methods for the company.

Companies are often reluctant to report such "minor thefts"

to the police. The theft did not involve just a mere stack of papers or a CD-ROM, laptop, or USB flash drive, however. It was the company's five-year strategic plan, the design of its next generation product, its pre-release last-quarter financial numbers. The value of the loss can easily run into millions of dollars. Once again, the police have resources for the apprehension and prosecution of thefts that is not available to the company that proceeds with just a civil lawsuit. In addition, if there is a criminal conviction, then liability will be established as a matter of law in the subsequent civil action and the only issue will be the amount of damages and the scope of injunctive relief. Even without a criminal conviction, juries expect theft incidents to be reported to police, and the failure to report the crime may impact the civil action if this fact becomes known to the jury.

Access by Hacking.

Access by hacking is the unauthorized access to information on the company's computers through their electronic connections to the outside world. Access to the company's computers by hacking is a criminal violation of the federal Computer Fraud and Abuse Act and often a criminal violation of the federal Economic Espionage Act, and the resulting access to proprietary information is an actionable misappropriation.

Access by hacking is the scenario that gives information technology managers nightmares. There is no method of access to the company's trade secrets that is as swift and as devastating as illegal access by hacking. The company can be fleeced without any knowledge at all that the incident occurred, and it can be accomplished through small cracks in an overall safe system. With high-speed connections in wide use by both companies and hackers, the company's entire portfolio of proprietary information can be downloaded in minutes once access is achieved.

Access by hacking has been the subject of many popular books and movies. Preventing, detecting, and documenting illegal access by hacking is its own professional discipline, with numerous digital security products now on the market. Which products are best suited for a particular company depends on many factors, including the resources of the firm, the magnitude of the threat, the downside risk of information loss, and the specific network, hardware, and software technologies utilized by the company. These topics are complex and beyond the scope of this book

The importance of a "defense in depth" should be stressed, however. The company should avoid a "Maginot Line" defense, with a very strong perimeter but little or no defense once the perimeter is breached. A hacker's access via the internet to the company's intranet or file server should not result in unprotected access to all of the company's proprietary information. In such a case, the theft of a single employee's dongle can result in the malicious download of the company's entire trade secret portfolio before the employee can even notice or report the theft.

An effective defense in depth against hacking will include many strategies. Proprietary information should be broken up so that all of the information required to replicate a specific technology isn't in one location. Access controls should limit every employee's access to his need-to-know information. Access tracking should create and archive records of access to proprietary information and retain them indefinitely. Highly sensitive information should not be accessible from outside the company's network at all. The most sensitive information should not be kept in any digital form, but retained only in paper copies on copy-proof paper in secure locations. The planning and implementation of these measures should be performed by experienced security professionals and carefully tailored to the company's needs and resources.

Access Through Inducement to Breach

Access through bribery or an inducement to breach an obligation of fiduciary or contractual confidentiality occurs when an outsider offers an insider some benefit as a quid pro quo for disclosing the company's proprietary information. In this case, the inducement to breach by the outsider is improper means, as is the breach itself — the disclosure of the proprietary information by the insider — and the resulting access to proprietary information is an actionable misappropriation.

The actual inducement may be in the form of cash, property, or other assets, or a favor done by the outsider for the insider. The inducement may also be in the form of a blackmail or extortion, in which the outsider threatens the insider with some unfavorable action if the insider does not disclose the proprietary information.

Since employees have the right to pursue other employment at any time, offering the insider a new job is not in itself inducement to breach unless the outsider makes it clear that the new job is a quid pro quo for the breach. An inference that the new job is a quid pro quo for disclosure of proprietary information may be made when the terms or compensation of the new position are inconsistent with industry or company practice. A 40% salary increase on the move to the new job, together with a $50,000 sign-on bonus for an engineer, for example, will lead the court to infer that the new employer induced the breach when a disclosure of proprietary information has been proven. For this reason, the terms of employment, including all compensation and benefits, of former employees with their new employers is frequently the subject of discovery requests in trade secret misappropriation lawsuits.

Prevention of access by inducement to breach begins in the company's hiring and promotion processes. Potential new employees being hired, and current employees being promoted, into sensitive positions with access to the company's proprietary information should be carefully screened. Prior criminal records, extravagant spending habits or heavy indebtedness, a history of substance abuse, and other risk factors that make the employee a likely target for inducement should be considered carefully before the hire or promotion is made. Employee education as to what their obligations of confidentiality are and the consequences of breaching those obligations is an important preventative measure, as is ensuring that employees have been properly informed as to what constitutes the company's proprietary information and is subject to an obligation of confidentiality.

Red Team Attacks

An effective way to evaluate a company's security against outsider threats is to employ the so-called "red team attack", named after the war games designation for military units playing the unfriendly forces. The red team attack consists of the company retaining outside competitive intelligence specialists to see if they can penetrate the company's security systems. The red team is retained by the company's general counsel and senior management without the knowledge of anyone else within the company. Without any inside information as to the company's security measures or the location or nature of its proprietary information, the red team sets out to penetrate the company's defenses and learn its trade secrets. The result of the red team attack is a report documenting what information the red team discovered and the means it used to discover it.

The results can be eye-opening, if not terrifying. The security of a company's proprietary information depends on its weakest link, and competitive intelligence agents specialize in discovering these weak links and exploiting them. By unleashing these experts against the company's own security systems, the company can identify its vulnerabilities and upgrade security measures where necessary

The use of red team attacks to test the company's defenses also has the beneficial effect of keeping employees on their security-conscious toes. System administrators benefit from practicing their system security skills against professional opposition in what amounts to a giant computer game. No one wants to be written up in a red team report as the weak link that resulted in a disclosure of the company's proprietary information, and certainly no one wants to be so written up twice.

Summation

Outsiders may gain access to the company's proprietary information through both proper and improper means. Access to trade secrets by proper means is always the result of employee carelessness, and the only effective prevention method is employee education. Access to trade secrets by proper means creates no actionable cause for the company, and the company's trade secret property rights in the information are destroyed. In contrast, access to trade secrets by improper means provides the company with the remedies of criminal prosecution for the improper act and civil actions for trade secret misappropriation. Red team attacks provide an effective means for testing the efficacy of the company's security systems against attacks by outsiders.

Chapter 7
Security Against Insiders

Executive Summary

The effectiveness of corporate security efforts aimed at outsider theft of proprietary information has now made insider theft the most common source of information loss. Losses to insiders can be lessened and their impact controlled through compartmentalization of information, access controls, and access tracking. Careful management of employee agreements and company ownership of all electronic devices used by the employee for handling proprietary information strengthen the company's legal position if a civil action becomes necessary. Some of these procedures also apply to contractors, consultants, suppliers, and customers who have insider access to trade secrets.

The Insider Threat

This chapter discusses protecting against theft of proprietary information by insiders. This is distinguished from the inadvertent and unprotected disclosures discussed in Chapter 4. Methods of protecting against inadvertent and unprotected disclosure with an effective trade secret culture will be discussed in Chapter 10.

It may not be immediately clear why any security against insiders is necessary. Insiders are the people who have access to proprietary information already, who have an obligation of confidentiality, either fiduciary or contractual, and who are already inside the fenced enclosure where trade secrets are developed, used, and stored. But that is precisely what makes the theft of proprietary information by insiders so easy. The "ASIS International 10th Trends in Proprietary Information Loss Survey" reports that the number of occurrences and total value of theft of proprietary information by insiders far exceeds that by outsiders.

Partly this is because the threat of theft of proprietary information by outsiders has been taken more seriously by companies, and the use of effective measures to reduce these losses has been growing. At the same time, companies shy away from security policies that may be interpreted as indicating that they do not trust their employees. In any case, the company's trade secrets are now more likely to be stolen by insiders than by outsiders.

This is not so surprising when one considers that, even with a low annual employee turnover of 15 percent, a company will have three people leave the firm each week for every thousand employees. These three departing employees are not leaving the firm because they are fulfilled, happy, and satisfied in their jobs. With the exception of retirees, they are leaving for potentially greener pastures. Their bonds of loyalty to the company are broken, and the company must rely on their integrity to honor their obligation of confidentiality.

Are most people employed by the company trustworthy? Of course. But a company of 15,000 employees with only 15% annual turnover loses forty-five employees a week. Are all forty-five of them trustworthy? Are all 180 employees who left last month trustworthy? Are all 2250 employees who left last year trustworthy? Not one bad apple in over two thousand people? Even among the people who were laid off or terminated for cause?

The company is faced with a seemingly insoluble problem. It must disclose its trade secrets to employees, contractors, consultants, and suppliers in order to obtain the competitive advantage that the trade secrets provide. Some small fraction of the employees, contractors, consultants, and suppliers to whom these trade secrets are entrusted can be expected to violate that trust and copy, use, or disclose those trade secrets for their own benefit or the benefit of others. And insider security precautions may be interpreted as a general distrust of employees, resulting in reduced employee morale, lower loyalty to the company, greater turnover, and a greater tendency on the part of employees to violate the company's trust and breach their obligation of confidentiality to the company.

The creation of a trade secret culture addresses this last concern. It will be discussed in a later chapter. This chapter will consider security against insiders, who must have access to trade secret information to perform their assigned duties.

Access Controls, Compartmentalization, and The Need to Know

While the company's trade secrets have to be disclosed to insiders in order for the company to use them, not all trade secrets have to be disclosed to every insider. The engineer working on the next generation product does not need to know the company's proprietary financial information or the confidential plans for the marketing campaign to field the product. The customer relationship manager does not need to know the latest laboratory results in the research department. The accounting clerk does not need to know the confidential introduction schedule for upcoming product features.

The disclosure of trade secrets to employees and other insiders should be limited to the disclosure of those trade secrets that the insider needs to know in order to further the business goals of the company. All other disclosures put the company's proprietary information at increased risk of insider theft.

Compartmentalization is one method for limiting the disclosure of proprietary information within the company. In compartmentalization, proprietary information is disclosed only within the areas of the company that require it in order to perform assigned tasks. Manufacturing trade secrets remain in manufacturing, engineering trade secrets remain in engineering, and so on. Most companies practice some compartmentalization, as it is unusual for financial trade secrets to be disclosed to company employees outside of the finance department and senior management. Broader and more effective compartmentalization is necessary to protect the company's trade secrets in other functional areas.

Even with functional-level compartmentalization, there remains a dangerously small cadre of employees who together possess all of the company's trade secrets. One former employee each from the company's finance, marketing, sales, engineering, and manufacturing departments may be enough to effectively reproduce the company's key technologies and processes. For this reason, further restrictions on disclosure beyond functional-level compartmentalization are often necessary.

Compartmentalization can also be implemented at the department, group, or project level. Which level is appropriate depends on the company's structure, the sensitivity of the information, and the risks of unauthorized use. In practice, different levels of compartmentalization within the company will be appropriate for different kinds of information, different functional areas, and different projects.

Compartmentalization is implemented using access controls. Access controls limit the access of employees to proprietary information on an individual basis. If the employee is an engineer in the Quality Assurance department working on Project X, he will be allowed access to engineering files that are compartmentalized on a functional level, Quality Assurance department files compartmentalized on the department level, and Project X files compartmentalized on the project level. He will not be allowed to access files from a different project, a different department, or a different functional area. If the employee is assigned to assist in the quality assurance of Project Y for a time, the access controls for his account can be modified to allow him access to the Project Y files during that period.

Access controls limit the damage resulting from the unauthorized copying, use, or disclosure of the company's proprietary information by limiting the information to which any individual or group of employees has access.

Access Tracking

Software for implementing access controls typically provides mechanisms for access tracking. Access tracking is important for proving access in any future trade secret misappropriation lawsuit that might be necessary to protect the company's proprietary information. Access tracking also provides a method of detection of unauthorized access and attempts at unauthorized access, as well as documenting unauthorized accesses and attempts.

Detection and documentation are important. If an employee

attempts to access company proprietary information to which he is not authorized, that person should be terminated for cause immediately. Attempting to access proprietary information beyond that which is necessary to perform his assigned work tasks is a serious danger to the company. Proprietary information exists to further the company's business goals. Attempts at unauthorized access are by definition contrary to the furtherance of the company's business goals and can only be intended to benefit others. Such unauthorized use of company proprietary information for personal benefit is by definition a breach of the employee's obligation of confidentiality. Unauthorized access — including access exceeding authorized access — to information contained on computers is also illegal under the federal Computer Fraud and Abuse Act.

Access tracking software should be configured to maintain logs of all accesses made to proprietary information. Since trade secrets may be retained indefinitely, logs that record accesses and activities on servers on which trade secrets are developed, used, and stored should be retained indefinitely.

Insider Access by Proper and Improper Means

The distinction was made in the previous chapter between outsider access to the company's proprietary information by proper and improper means, and we see now that the same distinction applies to insiders as well, with an important difference. Once proprietary information is compartmentalized and access controls implemented, the access to company proprietary information is divided into authorized and unauthorized access for each employee. The proprietary information to which the employees are authorized access is individually tailored to each employee based on their assigned work tasks. Access within the authorized set of proprietary information is proper, while access to proprietary information outside the authorized set is improper and actionable by the company, both as grounds for termination and legal action.

Insider access by proper means differs markedly from the outsider case, however. In the case of an outsider, the copy, disclosure, or use of proprietary information gained by proper means is not actionable, while the copy, disclosure, or use of proprietary information gained by improper means is a misappropriation. In contrast, the unauthorized copying, disclosure, or use of proprietary information by an insider constitutes a misappropriation regardless of how the information was accessed. When information is disclosed to an employee for a limited purpose such as carrying out assigned work tasks, exceeding that authorization constitutes a misappropriation even though the access was by proper means. Such conduct may also constitute a violation of the federal Computer Fraud and Abuse Act if it involves electronically-stored information.

The Employment Interview

The place to start protecting against the insider threat is at the initial employment interview. The prospective employee should be informed of the obligations he will undertake to his new employer to respect and protect intellectual property rights. His later acceptance of employment under these conditions is an explicit acknowledgement of the expected obligations to his new employer. The prospective employee should be provided a copy of the employment agreement at the interview, prior to the company making an offer and prior to the prospective employee informing his current employer that he is leaving the company. The courts often will not enforce the terms of an employment agreement presented to a new employee after he resigns from his former employment on equitable grounds.

It has already been pointed out that there is no "work made for hire" doctrine that applies to trade secrets, so the employment agreement should include a trade secrets assignment clause. The employment agreement should also contain a recitation of the employee's fiduciary obligation under law to maintain the confidentiality of the company's proprietary information. For senior employees or those in very sensitive positions, a non-compete clause may also be appropriate. The employment agreement should be carefully drafted by intellectual property counsel to conform to the state laws of the employee's work location.

Under no conditions should an employee who has failed to execute the employment agreement be allowed to start work or be permitted on the company's premises except as a visitor. No proprietary information of the company should be disclosed to the new employee until the employment agreement is executed. Until the employment agreement is executed, the new employee is a potential employee only.

Another reason to provide the employment agreement to the employee during the employment interview is that the potential employee has plenty of time to have an attorney review the document. It will speed up the process of getting all of the necessary paperwork executed on the first day of employment if the employee has had all of the documents reviewed in advance.

During Employment

A discussion of employee training and management on the subject of trade secrets will be discussed in the chapter on establishing a trade secret culture, but one recommendation should be made here. The employment agreement should be renewed annually during the employee's tenure with the company. In the event of a trade secret misappropriation lawsuit, the production of a twenty-year-old employment agreement dating back to the employee's first day on the job is not as compelling as twenty employment agreements signed by the employee each year during his employment.

The Exit Interview

One of the periods of highest risk of an employee copying, disclosing, or using trade secrets is immediately on his departure from the company. The proprietary information the employee knows is all current and valuable. The employee's bonds of loyalty to the company may be broken, especially if his departure is a layoff, termination for cause, or a resignation under unusual circumstances. The employee may be going to work for a direct competitor of the company, in the employ of whom the proprietary information would be immediately useful.

It is therefore recommended that the exit interview contain a trade secrets segment, the trade secret exit interview. During the trade secret exit interview, the employee is shown a high-level list of the categories of company trade secrets and proprietary information to which he has had access or been exposed during his employment. The employee is asked to sign a statement that acknowledges that 1) the company interviewer reviewed the employee's fiduciary obligation to protect the company's proprietary information, 2) the employee agrees not to disclose or use such information for his own benefit or the benefit of others, 3) the employee has returned to the company all copies of any proprietary information in his possession, 4) the employee does not have possession of any information relating to the proprietary information, and 5) the employee understands that if he has any doubt whether or not something is a trade secret, he is obligated to ask. The employee's certification should also be signed by the interviewer and both signatures witnessed by a third person. The trade secret exit interview certification should be retained indefinitely. A sample trade secret exit interview certification is provided in Appendix F.

What if the employee refuses to sign the trade secret exit interview certification? The certification merely restates the employee's pre-existing fiduciary obligation to protect the company's proprietary information and to return all such information to the company upon leaving employment. Signing the trade secret exit interview certification creates no new obligation. Failure to sign the trade secret exit interview certification, however, may be construed as evidence of bad intent in the event of a trade secret misappropriation lawsuit.

From the company's point of view, the trade secret exit interview protects the company from a later assertion of "I didn't understand my obligations" or "I didn't know this was a trade secret". The trade secret exit interview certification, signed by the interviewer and witness, constitutes contemporaneous evidence that the employee was placed on notice of his trade secret obligations upon his departure from the company, whether the employee signed the certification or not. The company will be able to strengthen its position that it has taken measures reasonable under the circumstances to protect its trade secrets.

Provision of Equipment

Today's mobile workforce relies on laptop computers, smart phones, USB flash drives, and a host of other portable electronic and digital storage devices. In many cases the employees use their own personal equipment while the company pays the service charges on expense reports. Permitting such use of personal devices is fraught with risk: the company should provide and pay for all electronic devices necessary for the employee's work tasks, and strictly prohibit the use of personal electronic devices for any company business. Up-front costs should not be a factor. The purchase of a laptop computer and a smart phone costs less than the fully burdened cost of even one day of the employee's time at most companies.

The inherent danger in allowing employees to use their own electronics lies in control of the devices when the employee leaves the company. When leaving the company, employees are required by law to return all company property, including any electronic devices they have been provided. The electronic devices and any proprietary information on them can thus be seized immediately by the company. In contrast, the company has no right to demand access to the employee's personal electronic devices upon termination of employment, even to ensure that company proprietary information has been deleted. Further, the inherent danger that the simple "deletion" of computer files will leave remnants of proprietary information on magnetic media that may later fall into the hands of outsiders has already been discussed.

The company has no effective way to ensure that its proprietary information on the employee's equipment has been completely deleted and is not being copied, used, or disclosed by the former employee. Trade secret lawsuits cannot be used as fishing expeditions to examine the personal computers of former employees absent evidence of other misconduct at the new employer. Permitting the use of personal computers and other electronic devices thus leaves a huge risk of loss unprotected when employees leave the company.

Non-Employee Insiders

Most of this chapter has been discussing employee insiders. With the exception of the discussion of employment agreements and exit interviews, most of it is equally applicable to third-party insiders like contractors, consultants, suppliers, and customers who obtain access to the company's proprietary information pursuant to the execution of a non-disclosure agreement. The insiders in this case have a contractual obligation of confidentiality to the company rather than the fiduciary obligation arising from the employer-employee relationship. Since the terms and conditions of the non-disclosure agreement are in the company's control, some recommendations relating to the non-disclosure agreement's content is in order.

A non-disclosure agreement should always specify that the proprietary information disclosed by the company is for a limited purpose. For a contractor, consultant, or supplier, the non-disclosure agreement should specify that the disclosure is for the limited purpose of the contractor, consultant, or supplier providing goods or services to the company. For the customer, the non-disclosure agreement should specify that the disclosure is for the limited purpose of customer evaluation or use of the company's products. Any other copying, disclosure, or use of the disclosed proprietary information is then an unauthorized copying, an unauthorized disclosure, or an unauthorized use. By definition, such other conduct thus constitutes trade secret misappropriation and is actionable.

In the case of contractors, consultants, and suppliers, all are providing goods and services to the company. The provision of such goods and services to the company should always be handled by the purchasing department and occur subsequent to the issuance of a purchase order. The standard non-disclosure agreement should be printed on the back of the company's purchase order form, with the following provision on the front of the form: "Acceptance of this order constitutes the acceptance of the terms and conditions printed on the back of this order." This is an effective way to create a non-disclosure obligation in every third-party relationship with the company.

Once again, since there is no "work made for hire" doctrine with regard to trade secrets, the non-disclosure agreement with contractors, consultants, and suppliers should contain a broad assignment of intellectual property rights to the company in all information developed by the contractors, consultants, and suppliers in the course of providing goods and services to the company. This assignment should be carefully drafted by intellectual property counsel to ensure that the company secures these rights to information that the company paid to have developed for its exclusive use.

Finally, contractors, consultants, and suppliers should be supplied with company-owned electronic devices necessary to perform their work tasks whenever possible. The return of these devices upon completion or termination of the contract should be specified in the contract terms and conditions. When it is not possible for the company to supply these devices, the contract should contain a right of inspection of all electronic devices used by the contractor, consultant, or supplier in providing goods and services to the company.

Summation

We have seen how the insider threat to proprietary information has now surpassed the outsider threat for most companies. This insider threat can be effectively managed by a combination of compartmentalization, access controls, and access tracking. It is also important to carefully manage employment agreements and ensure that all handling of proprietary information by employees is performed only on company-owned electronic devices. Similar measures apply to contractors, consultants, suppliers, and customers entrusted with insider access to the company's trade secrets.

Chapter 8
Inbound Security

Executive Summary

Significant financial losses can be incurred if the company is found liable for trade secret misappropriation by its employees. This inbound information leakage is most dangerous in the case of a new employee hired from a competitor. The company must have inbound security measures in place to ensure against this liability, including employee management policies and the compartmentalization and documentation of trade secret development. Inbound information leakage can also occur in the trade secret licensing process, resulting in the company being barred from independent development of the trade secret information if negotiations fail. Licensing negotiations can be structured to preserve the company's freedom to independently develop if negotiations fail, an alternate decision is made, or the information proves unsuitable to the company's needs.

The Inbound Security Threat

Even with airtight security of its own trade secrets, it is possible for the company to lose millions of dollars in civil litigation if it is found liable for a trade secret misappropriation committed by one of its employees. In this case, the proprietary information of another company is at issue. Recall the Uniform Trade Secret Act (UTSA) definition of misappropriation: "'Misappropriation' means ... disclosure or use of a trade secret of another without express or implied consent by a person who … at the time of disclosure or use, knew or had reason to know that his knowledge of the trade secret was … derived from or through a person who owed a duty to the person seeking relief to maintain its secrecy or limit its use…." Under the UTSA definition, if the company uses another company's trade secret without consent, and the company knew or had reason to know that the trade secret came from an employee who had a duty to the other company as a former employee to maintain the secrecy of the information, then the company is guilty of trade secret misappropriation.

The key phrase in the UTSA definition is "knew or had reason to know", but some courts have recently applied the strict liability doctrine of *respondeat superior* in employee-employer relationships. The company can be found vicariously liable for a trade secret misappropriation under the doctrine of *respondeat superior* ("let the master answer") if it can be shown that (1) the company had direct control over the employee's actions and (2) the company was the beneficiary of the misappropriation. Both of these conditions are likely to be true in the case of trade secret misappropriation by an employee of the company. The employee is assumed to be under the direction of the employer in performing his assigned work tasks, and the economic benefit of a trade secret misappropriation in performing those tasks inevitably accrues to the company.

Avoidance of liability under *respondeat superior* requires reasonable measures to protect against the importation of trade secrets from others. Since the employee was under the company's direct control, the company should have taken reasonable measures to ensure that the employee did not commit a tortious act as its agent. Thus a "look the other way" or "see no evil" policy with regard to employee misappropriation of another's proprietary information is precisely the wrong policy with regard to inbound misappropriation. The "ostrich defense" will not shield the company from liability in a trade secret misappropriation lawsuit.

The consequences of an adverse judgment with regard to inbound trade secret misappropriation, whether under the *respondeat superior* doctrine or under the UTSA's "knew or had reason to know" provision, can be severe. Courts often order injunctions that prevent the misappropriating company from any further use of the tainted information. Prior to the misappropriation, the company was free to independently develop the same trade secret. After an adverse judgment in a trade secret misappropriation lawsuit, independent development is often no longer possible because the company has been exposed to the purloined trade secret information. The only effective remedy available to the court is a non-use injunction that prevents the company from using the trade secret information for the period of time necessary to eliminate any lead time advantage obtained from exposure to the trade secret information.

Employee Management

The first step in managing the inbound security threat is to properly manage employees. It is imperative that the company make it clear in writing to all employees, both at the time of hire and periodically thereafter, that no trade secret information to which they may have been exposed in prior employment is to be disclosed or used within the company under any circumstances. New employees should be required to sign an acknowledgement of this company policy and the certification maintained in the employee's personnel file. The company should provide all employees with contact information with attorneys within the company's legal department and direct the employees to discuss with these attorneys any concerns that may arise in this regard during the course of their employment.

A high-risk situation arises when an employee receives an assignment that would make the proprietary information of a former employer immediately useful in the performance of the work. The employee cannot "unlearn" what he knows. He already knows which approaches will not work and what the correct solution is, information that was developed at the expense of his former employer. To allow the employee to work on this assignment raises the specter of the inevitable disclosure of the former employer's trade secrets, and can result in substantial liability for the company in a trade secret misappropriation lawsuit filed by the former employer. The safe course of action in these circumstances is to assign the employee to other tasks that reduce the risk of a lawsuit by the former employer. As only the former employee knows what the former employer's trade secrets are, it is the employee's responsibility to bring these situations to the attention of the company.

Documentation and Compartmentalization

Thorough documentation and effective compartmentalization of proprietary information help to protect the company from trade secret misappropriation claims. If the company's employees freely discuss the development of trade secret information across project boundaries, and documentation of the development of trade secret information is haphazard or nonexistent, it may be impossible to prove that the information was independently developed. Fending off allegations of improper use of another's trade secrets will be even more difficult without effective procedures and thorough documentation, including dates and times, in place.

If the development and discussions of trade secret information within the company are compartmentalized and well documented, however, defenses against alleged misappropriation are bolstered. Compartmentalization may show that the plaintiff's former employees never worked on the project in question, never had access to the proprietary information in question, and never participated in any discussions about the project. Thorough documentation of research and development projects can show that the information was developed independently by detailing the work activities and results of the intermediate steps of the independent development process.

Effective compartmentalization and thorough documentation of research, development, and internal discussions of trade secrets offer other significant benefits to the company, which we discuss elsewhere, but their importance in protecting the company from alleged trade secret misappropriation claims by third parties in itself justifies their implementation.

The New Employee

The biggest danger of an actual inbound trade secret theft occurs in the case of a new employee hired from a direct head-to-head competitor. The information is current and directly applicable to projects underway at the new employer, and the ethics and conduct of the new employee are not yet known. There are some warning signs that managers should look for in managing new employees.

While people often talk about someone "hitting the ground running", in fact it usually takes some time for new employees to become productive. The new engineer, the new salesman, and the new product manager need time to learn the company's technology, its products, and its markets and to effectively apply their general knowledge, skills, and experience to the company's business. If a new engineer records several patentable ideas in his engineering notebook in his first week at the new company, if a new salesman lands a major new client in his first month representing the company, or if a new product manager quickly devises a complex plan for penetrating a tough market, then the new employer faces substantial risks. These high-risk situations should be monitored by management and intellectual property counsel.

Inbound Security in the Licensing of Trade Secrets

Because it is legal to independently develop trade secret information, the licensing of trade secrets is necessarily different than the licensing of patents. In the case of patent licensing negotiations, the patented invention has already been disclosed. In contrast, in trade secret licensing negotiations, the trade secret information has never been disclosed. It is often in the company's best interests not to obtain the full details of the trade secret until after the licensing agreement is executed.

Consider the company faced with the "make-or-buy" decision for a trade secret asset. The company could spend significant time, effort, and money attempting to develop the trade secret solution it needs to solve a certain problem, or it could license the trade secret solution from a third party and pay substantial royalties. The decision will be based on the computed costs and lead time to develop, as well as such considerations as whether the technology is a key product differentiator that should be owned and controlled by the company. But if the information is disclosed under a non-disclosure agreement during the negotiations, the company will be contractually bound not to use the information if the licensing negotiations fall through. The company will no longer have the option of independent development of the same information. Even worse, the company will not really know what information it will be barred from independently developing until the information is disclosed.

There's a further complication. Since licensing negotiations are conducted without full disclosure of the information, the company may find after licensing the trade secret information that the information is not that useful to the company. For this reason, licensing of a trade secret should be on a royalty basis, in which licensing fees are calculated based on the extent of use by the licensee. The license agreement should also include opt-out provisions if the trade secret information becomes generally known or is obtained from another source without an obligation of confidentiality.

One method for conducting more informed licensing negotiations without barring the company from potential independent development is to use a third-party expert to examine the trade secret information for application to the company's business. The trade secret information is disclosed under a non-disclosure agreement to this third-party expert, who makes the necessary evaluations to assist the company in its licensing decision. If the evaluation is done internally, another option is to require "Chinese wall" provisions to protect the trade secret information from disclosure to others in the company in the event that a licensing agreement is not consummated.

Summation

Security of the company's own trade secret information is not enough to protect the company from adverse financial consequences with respect to trade secret assets. The company must also protect against the leakage into the company of trade secret information belonging to others. This inbound information leakage or exposure can occur improperly, through the actions of employees, and the company can be held liable. Employee management processes must be in place to ensure against such liabilities, while compartmentalization and documentation of trade secret development should be used to protect against the company being unjustly accused of inbound misappropriation. Inbound proprietary information leakage that bars the company from independent development can also occur in negotiations to license trade secret information. The trade secret licensing process should be structured to maintain the company's freedom to independently develop the trade secret information if negotiations fail, an alternate decision is reached, or the trade secret information ultimately proves unsuitable to the company's needs.

Chapter 9
Monitoring

Executive Summary

Because trade secret information assets can be stolen while they remain in the company's possession, a trade secret theft may not be detected until the information appears in the marketplace. Companies must monitor the Internet, their business environment, and their internal networks to ensure that proprietary information does not appear outside of authorized locations. If proprietary information is discovered outside of authorized locations, investigation is warranted to determine if and how information leaks occurred. An internal investigation may determine whether the proprietary information was misappropriated or independently developed. Sometimes civil litigation will be necessary to discover the truth and provide evidence of misappropriation.

Detection of Information Theft Through Monitoring

Many thefts of trade secret information go undetected, but evidence of the theft surfaces later. It is important to proactively monitor the company's business environment to detect the theft of proprietary information so that corrective measures can be taken to address the situation before it gets worse. In many cases it will be possible to avert the damage or to keep further losses from occurring.

The first step is to ascertain whether the company's trade secret information exists outside the company. The second step is to determine whether such outside information was independently developed or whether it came from within the company through an insider or outsider disclosure. The third step is to take legal action if there is no evidence to support the independent development of the information.

Monitoring the Internet

The wide availability of search engines, both externally such as Google and internally such as web crawler software, allows the company to search the Internet for the presence of trade secret leaks on web sites and newsgroups. Resume sites such as Monster.com should also be searched for unauthorized disclosures of company proprietary information.

One caution is in order with regard to searches using external, as opposed to internal, web-crawler software and search engines. Although Google and other search service providers have been very protective of user search data, the company should not use search terms that in fact disclose the trade secret information when using outside search engines. Such searches could constitute an unprotected disclosure and result in forfeiture of trade secret rights in the information. Search terms should be crafted to turn up the information without disclosing it.

Monitoring the Business Environment

It is also important to monitor the company's business environment. New product introductions, changes in pricing and features, and customer losses to competitors may signal proprietary information leakage. Monitoring of the business environment should include review of competitor web sites, careful attention to industry news in both hard-copy and on-line sources, attendance at industry shows and conferences with specific attention paid to competitor offerings, and tracking of key customer accounts. Changes in the competitive outlook in these areas should be analyzed for potential leakage of the company's proprietary information.

The introduction of a new product or feature, or a significant price reduction, by a competitor may correlate with the past loss of one or more key engineering or research employees to the competitor, employees who were in a position to know company proprietary information that would make such an introduction or price reduction possible. The loss of key customers to a competitor may correlate with the past loss of one or more key sales or marketing people to the competitor, employees who had access to company proprietary information that would make such an account takeover possible. These situations require investigation to determine whether there is circumstantial evidence to support a trade secret misappropriation action.

Monitoring the Intranet

It is also important to monitor the company's internal network. In this case, company proprietary information is assumed to be present. The necessary inquiry in this context is whether the information is where it is supposed to be, or, more precisely, is it where it is not supposed to be.

Compartmentalization, access control, and access tracking are important methods in protecting against the threat of theft of proprietary information by insiders. Trade secret information located within the company outside of approved locations defeats all three of these methods. Internal searches that show proprietary information outside of approved locations indicate either trade secret misappropriation or failure to comply with the company's information protection measures. Such sidestepping of the company's information protection measures is more likely to occur when the importance of such measures has not been effectively communicated to employees, and as a consequence the employees see these measures as an impediment to getting their work done.

The field of information policy enforcement software is rapidly developing, with new products and companies entering the marketplace. Such regimes can be helpful in measuring compliance with company programs, but it is ultimately employee buy-in that will determine the success of a company's information protection program.

Independent Development or Misappropriation?

Once it has been discovered that the company's proprietary information has been compromised in some way, the question becomes whether the acquisition of the information was due to an actionable misappropriation, legal independent development by a third party, or a failure on the company's part to take reasonable measures to protect the information. One of the key issues in answering this question is access. There cannot be independent development if one or more of the developers had prior access to the trade secret information. An investigation is necessary to determine if the third party was exposed to the company's proprietary information.

Access tracking by the proprietary information holder is a powerful way to determine who had access to the company's information and when such access occurred. Access tracking will limit the number of possible sources for information leaks and streamline the investigation into the source of the leaks. If the information has been compartmentalized, and the accesses to the information tracked and logged, the investigation starts with a much shorter list of possible suspects.

The transmission path must then be determined. Did one of the employees who had access to the information leave the company and enter into the employ of the third party? Was the timing of their departure consistent with such a misappropriation? If the competitor's new product came out six months after the employee left the company and entered into employment with the competitor, the timing supports a preliminary conclusion that a misappropriation occurred. If they came out with the new product a week after his departure, such a conclusion may be misplaced.

If a credible case for access and transmission can be made, filing suit for trade secret misappropriation against both the former employee and the third-party company provides the opportunity for discovery under the applicable rules of civil procedure to uncover further evidence of wrongdoing. Often, the internal investigation is just the tip of the iceberg, and formal discovery will ultimately reveal the full scope of the trade secret misappropriation.

Summation

Trade secret misappropriation is often not discovered until evidence of the company's proprietary information appears in the marketplace. The company should monitor the Internet, its business environment, and its intranet to ensure that proprietary information has not wandered out of authorized locations. Once proprietary information is discovered in an inappropriate location, further investigation is required to determine whether the information was misappropriated or independently developed. The commencement of civil litigation for trade secret misappropriation is often required to resolve these contested issues.

Chapter 10
Suspect Profiles

Executive Summary

Misappropriators of trade secrets are highly intelligent, motivated individuals who often fit a rather specific suspect profile. They are not the secretaries, janitors, and technicians of fiction, but the senior technical and management personnel who understand the technology they are stealing. Domestic and foreign misappropriator subject profiles can be helpful in providing guidance as to where additional investigation may be most helpful.

Methodology

Anyone who has watched crime investigation programs on television has been exposed to the concepts of evidence collection, forensics, and suspect profiling. We wondered if we could use our experience in trade secret misappropriation cases to draw up a suspect profile for the trade secret misappropriator. Could we in fact determine a profile of an information thief? Although not a scientific study by any means, what we started on a lark soon became very interesting.

We began by writing a separate description for each domestic trade secret misappropriator we had run across in our trade secret litigation matters. We left out cases for which we did not have enough data to draw up a complete description, as well as cases in which we were unsure from the evidence whether a misappropriation had occurred. As many cases settle outside of court, there may be no finding of fact upon which to rely. In these cases, we used our best judgment from an analysis of the case to determine whether to include the alleged misappropriator in our private study or not.

We then compared the descriptions across our trade secret litigation matters to determine if there were any descriptors that were common across cases and which could form the basis for a suspect profile. We found that all of the misappropriator descriptions looked like they could be descriptions *of a single individual.* The amount of common traits across our known universe of misappropriators was quite a shock.

Findings for Domestic Misappropriation

Our findings are shown in Table 1. While some of the findings are perhaps to be expected, others are counter-intuitive or even a bit surprising. "White male, 25 to 45 years old" was something we expected, but that all the misappropriators were married men, usually with children, very intelligent, and long-time employees surprised us.

Table 10-1. Domestic Trade Secrets Misappropriator Profile

White male, 25 - 45 years old
Married, with children
Intelligent
High self-esteem/arrogant
Under-educated for his position
Well compensated
Long time with the company
Learned the technology while at the company
Self-starter

Independent worker
Made highly significant contributions
Tends to think of/describe work performed as "his plan", "his software", "his idea"
Alienated from company management
Feelings of being under-appreciated; chip on his shoulder
Starts competitor or goes to competitor where friends from the current employer are
Steals product technology, but also steals customer and vendor info
Motive: money, not for greed, but for validation

Psychology of a Domestic Information Thief

The consistent theme through all of these cases was the psychology of the misappropriator. Each of these individuals was a self-starting, independent worker who had worked his way up through the company to a responsible, well-compensated position. Each was highly intelligent, and often felt their intelligence was more important than, and a substitute for, experience. They had generally exceeded the level in the company that the level of their education would best match. That is, they were doing PhD-level work with a Bachelor's or Master's degree, were doing work one would expect of a college graduate with only a high school education or GED, etc. They were all driven and impatient: too impatient to get the degree or advanced degree, and impatient for promotion.

Each had made significant contributions to the company, and was highly regarded for those contributions, often being the guru or go-to guy in their area of expertise. All felt under-appreciated, and that management did not understand how much their work product had contributed to the bottom line.

Many felt that the company's revenues, or, in larger companies, its revenues in a specific product line, were completely dependent on these contributions. In most cases, that was at least partially true. With this mindset, any perceived slight, such as the promotion of another instead of them, or a lower than expected amount on a bonus, could become a triggering event for a building resentment and eventual misappropriation.

The misappropriators also viewed their work product as being their property, despite their statutory employee status and appropriate invention agreements. As these contributions, in their view, drove the company's revenues, they also looked at the company's customer and vendor lists, which often included customers and vendors they had helped secure, as being fair game to take with them as well.

They felt that company information based on their work product was theirs to take with them to their new employment, which was always with either a startup or an existing competitor for whom other former employees also worked. Thus they moved to an environment where they already had the respect of their peers and did not have to once again earn the respect of co-workers. Their sometimes greatly increased salary validated them and proved to them how poorly appreciated they were in the former firm.

While each of these individuals felt that their work product was theirs, and they should be free morally to use it, all took efforts to conceal the theft. All tried to conceal the identity of their new employer from the current employer when resigning. Due to their intelligence and self esteem – or downright arrogance – all felt they had sufficiently covered their tracks to get away with the theft, yet their efforts were usually laughably insufficient to conceal the theft from straightforward computer forensics and other evidence-gathering techniques. Like most insurance fraudsters, they all thought that in this area, in which they were rank amateurs, their superior intelligence would be sufficient to outsmart trained investigators, whom they viewed with intellectual contempt.

All vigorously protested their innocence, and were not above perjury in denying incriminating actions that were later well-documented by the evidence. They acted as though their actions were completely justified, and the rules by which others lived did not apply to them. All acted smug and superior when appearing for deposition, although a vigorous deposition based on the discovered evidence could leave them pale and shaken.

The new employers also vigorously denied any wrongdoing. While we may suspect that in many cases there was after-the-fact knowledge that misappropriated information was brought into the company, or even that the misappropriated information was one of the reasons for the initial hire of the misappropriator (thus justifying the greatly increased salary), these suspicions are seldom proven.

What We Didn't Find

Curiously, we did not see many misappropriations in our sample committed by women, nor by men over the age of 45 years, nor by minorities. The particular mindset required to justify domestic misappropriation seems peculiar to white males between 25 and 45 years old.

Additionally, we did not see misappropriations of technical information by technician-level employees, which is one of the groups about which employers often have concern, nor by secretaries, janitors, cleaning people, etc.. The information was instead misappropriated by trusted higher-level engineers, researchers, and executives most able to make use of it in their future employment.

Findings for International Misappropriation

The above profile only applies to domestic trade secret misappropriation. We therefore thought it would be worthwhile to pass on what we have learned about the international misappropriator in studying such cases.

Table 10-2. International Trade Secrets Misappropriator Profile

Born in a foreign country
May be either naturalized U.S. citizen or foreign national on a work or student visa
May be male or female, married or single, with or without children, of almost any age
Often a group effort, such as a husband and wife team
Usually from China, France, Israel, India, Japan, or Russia
Has close family remaining in the birth country, often parents
Retains close ties to the birth country, traveling there often
Intelligent
Highly educated, often with a PhD. in a technical field
The theft is sponsored by a government agency, university, or technology company in the birth country
Ultimately will return to the birth country and work for the sponsor
Steals primarily product technology
Theft may be through a front company, which shields the

sponsor
Motive: nationalism

Our findings are shown in Table 2. Again, many of the findings are not surprising: foreign nationals or naturalized citizens with strong ties to their birth country were traits we expected. The very high levels of education were at first surprising, but on second thought it made sense: the sponsor would pick agents who would be best equipped to get a job in a high-tech firm, best equipped to determine which technologies to steal, and best equipped to minimize the amount of documentation that needed to be transferred, carrying the rest in their heads. What was most surprising was the number of women involved, the use of groups, and the use of husband and wife teams, often with one inside the target company and one outside working at a front company.

Psychology of an International Information Thief

We do not have enough case data nor enough information on particular misappropriators to be nearly as detailed in our description of the psychology of the international misappropriator as we were with the domestic misappropriator. Nevertheless, some discussion is possible.

With the exception of Israel, each of the most likely birth countries has a long and proud history, including an imperial period. With the exception of Israel and France, each lagged the technical developments of Western countries well into the 20th century; these have been playing a "catch-up" ball game. Israel has been involved in an existential struggle since its founding.

All of the most likely birth countries engender unusually strong ties of national loyalty in their expatriates. All of the most likely birth countries have weaker legal protections for trade secrets than the United States. Business competition in most of the most likely birth countries is more rough and tumble than it is in the United States, where corporate competition is more of a "gentleman's game", with its own rules and etiquette.

Further, if one's parents and siblings, uncles and aunts, nieces, nephews, and cousins live in a country where the living wage is a fraction of what it is in the United States, where the life expectancy is a decade or more shorter than in the United States, where malnutrition is commonplace and adequate healthcare is rare, trade secret misappropriation may not be the injustice that ranks highest in one's mind. It must be particularly galling to watch children in the birth country work 12-hour days for pennies without complaint and then travel to the United States where children complain bitterly if their parents limit their cellular data plan. The desire to assist the birth country in catching up to the West in critical quality of life areas is clearly a strong motivator.

The international trade secret misappropriator has, then, in his own mind, the moral high ground. Theft of trade secrets, perjury, obstruction of justice and spoliation of evidence to cover the theft, and fleeing criminal prosecution or civil penalties are all morally justified and of little consequence in this mindset. And, unlike the domestic trade secret misappropriator, they always have the birth country to go home to.

While this point of view can be understood, we cannot condone the theft of trade secret information by international misappropriators, whose sponsors then market their products internationally in competition with our own firms where the technology originated. But understanding the mindset of the misappropriator is critical to a successful effort to secure our trade secrets.

One final note on the international trade secret misappropriator: the last trip home. In most of these cases, the misappropriator makes a trip home "to visit their parents," as they have often done before. They then return to the United States, often calling in sick after as little as a day. And then they are gone, back to the birth country. Analysis of the evidence leads us to believe the last trip home is to deliver the proceeds of the theft for analysis by their sponsor. The final short trip back to the United States is to "fill in the gaps". A massive data dump to USB flash drives is made in that short return period. Finally, calling in sick gives them time to get out of the country before it becomes clear that they are not coming back to work.

Another common element is taking leave "for medical reasons". The misappropriator might leave the target firm for a month, or two, or even as much as a year, and then return to employment. This return to work is usually very short, on the order of days. The truth is that the leave had nothing to do with medical reasons unless, perhaps, one includes homesickness. The misappropriator returned to the birth country, where they obtained employment in the same field. They subsequently return to the United States for a short period to clean out the technology files of the target company, their former employer, and return almost immediately to the birth country and their new employer.

Finally, front companies are often used. The technology is stolen by insiders working for the benefit of the front company and its sponsor. The technology is then embedded in a new product, and sold or licensed to the sponsor by the front company. These thefts can go on for years, with people inside and outside the target company conspiring to steal new technology as it is developed. The internal connections are so good that the sponsor can often beat the target company to market with the target company's own technology.

A Few Words Of Caution

As anyone who has watched the criminal investigation programs knows, suspect profiles must be used with care. The profiles reflect the typical offender for the type of crime considered, and are only to be used by investigators as hints about where greater efforts may pay off. In particular, they shouldn't be used to eliminate suspects.

So too here, our profile of a trade secret misappropriator is intended as a guide to further investigation when a misappropriation is suspected. It should be used neither as exculpatory nor incriminating evidence, but merely as a bit of a hint about where further investigation might most likely reap results.

Further, our sample size is not large. We considered domestic trade secret misappropriation cases with which we were sufficiently familiar to be able to compile a detailed description. Our own practices may not reflect the universe of trade secret cases and this further biases the sample. For the international misappropriator, we relied on what court documents were released as well as reports in the trade and general press. Many documents in trade secret cases are filed and remain under seal, and this biases our sample as well.

In particular, we are not able, in cases in which we were not directly involved, nor allowed, in cases in which we were directly involved, to consider information that is not available in the public record. The authors take court-ordered protections on confidential information very seriously, and could include no information, or derivatives thereof, that are subject to court-ordered protections in researching or writing this chapter or the book as a whole.

Finally, neither of the authors is a trained psychologist, and our opinions on the profile and psychology of the trade secret misappropriator must be considered the lay opinions of those experienced with trade secret litigation. Our combined fifty years of experience in trade secrets cases provides, we believe, valuable insight on this topic, but should not be considered by any means an expert opinion.

The FBI Warning Signs

The Federal Bureau of Investigation has also offered warning signs of a possible misappropriator in a press release dated May 11, 2012. Their warning signs are all behaviors of the misappropriator, rather than a profile, but they fit well in this chapter and echo some of our own observations.

> Here are some warning signs that *could* indicate that employees are spying and/or stealing secrets from their company:
>
> - They work odd hours without authorization.
> - Without need or authorization, they take proprietary or other information home in hard copy form and/or on thumb drives, computer disks, or e-mail.
> - They unnecessarily copy material, especially if it's proprietary or classified.
> - They disregard company policies about installing personal software or hardware, accessing restricted websites, conducting unauthorized searches, or downloading confidential material.

- They take short trips to foreign countries for unexplained reasons.
- They engage in suspicious personal contacts with competitors, business partners, or other unauthorized individuals.
- They buy things they can't afford.
- They are overwhelmed by life crises or career disappointments.
- They are concerned about being investigated, leaving traps to detect searches of their home or office or looking for listening devices or cameras.

Summation

The domestic trade secret misappropriator often fits a particular suspect profile and its accompanying psychology, while the international trade secret misappropriator is both more varied in profile and more particular in circumstance. These profiles can be used by investigators in trade secret misappropriation cases as a hint about where further investigation might be most rewarding.

We believe that further study of these profiles by those with access to greater sample sizes would be of benefit in further understanding the nature of trade secret misappropriation and in protecting the confidential and proprietary information of U.S. companies.

Chapter 11
Establishing a Trade Secret Culture

Executive Summary

There is no silver bullet for proprietary information protection, no hardware widget or software program or canned process that can provide proprietary information protection in the absence of employee involvement. A corporate trade secret culture requires both top-down communication of goals and bottom-up communication of needs to be effective. Management must provide the necessary tools — equipment, process, and training — for effective proprietary information protection, and involve employees in the selection and use of the best suite of tools for the company's information protection program.

The Importance of a Trade Secret Culture

Many of the recommendations in the security chapters of this book — physical security, data security, legal security, vendor management, compartmentalization, access controls, access tracking, and monitoring of internal and external networks — are the focus of new information technology products now appearing in the marketplace. It seems that we hear about new companies and new products in these areas on a weekly basis. But no information technology can provide the company with proprietary information security in the absence of the cooperation and active involvement of the company's employees.

Successful trade secret protection programs enlist the aid of all the employees in the company. It is not an effective option to name a few employees as a trade secret protection group, or to install some new information security product, while the rest of the employees continue business as usual. This is because once a trade secret is divulged in a careless, inadvertent, or unprotected manner by any employee, the trade secret rights are lost. The company's trade secret protection therefore depends on the weakest link in the chain. Management may wish for a simple technological solution to protect trade secrets, but no such product exists.

The easiest way to enlist the support of all the company's employees is to work toward a culture of trade secret awareness. Such a cultural awareness is much more effective than standalone employee training sessions, and it is ultimately more economical. Employee training as a standalone activity has to be renewed periodically, while a healthy trade secret culture is more self-sustaining. Employee training — especially new-employee training — fades in memory with time, while a healthy trade secret culture will reinforce itself. New employees will pick up the trade secret culture along with the rest of the corporate culture more completely than anything that can be done in a few minutes of a new-hire orientation.

Top-Down Communication

For a trade secret culture to be effective, management must effectively communicate its goals to employees. This requires much more than a "Here is the new proprietary information policy, read it, sign here, and return it" approach. Employees must understand that the company's trade secrets and proprietary information are the source of the company's competitive advantage. They need to know that risks to the company's trade secret information are risks to the company's revenues, earnings, and share price, and ultimately risks to the employees' jobs. Management's efforts to protect the company's trade secrets from misappropriation or loss are efforts to protect the employees' jobs, their stock options, and their pensions. If employees understand this, company information security policies will be accepted and followed. Absent this understanding, all the company policy manuals and policy communications in the world will be worthless.

The most important part of top-down information security communications, then, is not training about how to use this security widget or follow that security policy. The most important part of top-down information security communications is the effective communication of management's goals to protect revenues, earnings, share price, and jobs. If the employees understand management's goals, they will figure out how to use the new security software, or ask for further training to use it, or have someone help them learn how to use it. No one will have to enforce attendance at training sessions for the new security widget when employees are motivated to use it.

The other important aspect of top-down information security communications is to make sure that employees understand that management wants to hear any and all employee suggestions about ways to improve the company's proprietary information security. No matter how much effort, money, and consultant time is put into information security, employees involved in the company's day-to-day business operations will find gaps in the information security program that management and outsiders cannot. To close those security holes, management needs to communicate to employees that information on those gaps and how to close them is earnestly desired. This cannot be empty promises. Management has to convince employees that it is open to bottom-up communications on information security, and mean it. Management at all levels must walk the talk on this.

Bottom-Up Communications

There are several types of bottom-up communications that must be a part of an effective trade secret culture. First, employees need to have a means of communicating information about gaps in the company's information security program, as well as suggestions on how to close them. As these gaps may involve other employees or groups within the company, these communications need to be confidential, and employees must know that they are confidential. Financial incentives for unearthing the largest information security gaps are in order and will encourage participation.

Employees also need a way to quickly and easily communicate the detection of company proprietary information in unauthorized locations. We have already discussed the need to monitor the Internet, the company's business environment, and the company's internal networks. In actual cases, it is usually employees who have knowledge of the company's proprietary information who first notice company proprietary information in unauthorized locations. Immediate communication of this detection to the company via a 24-hour toll-free number is the fastest way to get this information to the appropriate people within the company so that a timely response can be made. This number can also be used to quickly report the theft of company or employee property containing trade secret information, such as laptop computers and briefcases.

While most such detections will involve proprietary information in the hands of outside third parties, some will involve the improper access or storage of proprietary information by another employee. Thus this reporting too must be confidential, and understood by the employees to be confidential. Once again, financial incentives for detecting the most important proprietary information leaks are in order and will encourage participation.

Finally, employees need a way to request assistance with information security issues. Employee agreements usually include clauses that employees will respect the confidentiality of the company's trade secrets and proprietary information, but seldom include any workable definitions of what constitutes the company's trade secrets and proprietary information. How is the employee to know? When an employee is preparing a paper for presentation at an industry conference, who does he ask to clarify what is permissible to discuss in his paper and what is not? The leakage of proprietary information through employee resumes has already been discussed. When an employee prepares his resume, how does he determine what can and cannot be on the resume? Employees need a way to request confidential assistance in addressing proprietary information concerns in the workplace.

Enablement

In order for a trade secret culture to work, employees need the tools to implement an effective information security program. Even the most motivated employees in a trade secret culture cannot implement an information protection program without the proper tools. These tools fall into three main categories: equipment, process, and training.

Information security equipment includes software that provides document rights management, encryption, access controls, access tracking, firewalls, hacking detection, and other benefits in addition to physical hardware. The software product choices in this area, and the benefits and features they provide, are expanding rapidly, and any package chosen now should be considered an interim solution to be reviewed annually.

Information security equipment also includes more conventional items such as card access systems, security camera systems, security glass, fences, safes, locking document storage cabinets, and other physical access controls and monitoring, as well as document disposal bins and document shredders.

Information security process includes: company policies; employee agreements and non-disclosure agreements; document storage, copying, retention, and destruction procedures; pre-release review of public information releases, including conference papers, press releases, and corporate web pages; compartmentalization and access control policies; and incident reporting policies.

Information security training should include training about how to identify the company's proprietary information and the employees' fiduciary obligations to maintain its confidentiality, as well as training in the use of information security equipment and the execution of the company's information security process.

Employees from across the company should be involved in the selection and configuration of information security equipment, the design of information security process, and the scheduling and content of information security training. Once the employees understand and buy into the need for proprietary information security, they represent the company's best resource in designing an effective information security program. Outside experts should be brought in to assist in tailoring an information security solution, but the employees who use the company's trade secrets in the day-to-day business of the company know the company's processes and needs better than any outside expert. Involving them early, and keeping them involved, ensures a better information security solution and continued employee support for the program.

Summation

The employees are the essential foundation of an effective trade secret protection program. Enlisting employees as active participants in the process requires an unambiguous and committed top-down communication of goals as well as an effective bottom-up communications channel of needs. The selection and use of a suite of tools for trade secret information protection should involve the employees early, and keep them involved, both to ensure the best program possible for the company's needs and to maintain employee support for the program.

PART III: ACCOUNTING

[I]t appears that you may have acquired additional intangible assets from this acquisition. Such additional identifiable intangible assets appear to be licensing agreements, non-compete agreements and trade secrets and other technical information. Please provide us with copies of the independent appraisal reports used to determine the fair value of your intangible assets when you allocated the purchase price for the company. Also, please tell us why you have not included a fair value for licensing agreements, non-compete agreements or trade secrets and other technical information....

— SEC enforcement letter

Chapter 12
Inventory and Classification

Executive Summary

Accounting for trade secrets requires an inventory of the company's trade secret portfolio. Creating a useful inventory of the company's trade secret portfolio entails four steps: inventorying the potential trade secrets, categorizing the potential trade secrets, identifying the actual trade secrets among the potential ones, and classifying the actual trade secrets. All of this information must be obtained from employees who know what the company's trade secrets are, either directly from the employees, through the use of consultants to interview the employees, or by use of automated methods to collect information from the employees. The information collected from employees needs to be reviewed by skilled evaluators to generate an accurate and useful inventory. The inventory must be updated periodically to reflect changes in the company's dynamic trade secret portfolio.

Inventory of Potential Trade Secrets

The first step in accounting for any group of assets is to take an inventory of the assets. The company must have a list of what the assets are before any further steps such as classification, valuation, and reporting can be taken. Inventorying trade secrets is difficult because information assets seldom have a physical presence, and they are being continuously created and destroyed — classified, reclassified, and declassified — in the normal business operations of the company. As trade secrets are seldom purchased, there is no record of invoices that can be used to generate a list of trade secrets. There is no physical location that can be inventoried for trade secrets. The company's entire trade secret portfolio is an amorphous, intangible, and inchoate cloud of information, stored on paper, in computer drives, and in the minds of employees.

The key to this puzzle is that any trade secrets used in the normal course of business — any information that derives value from being secret — is generally known to some employee of the company. The employees, as a group, know what the company's trade secrets are, or, more precisely, they know all the information that may qualify as a trade secret. The employees often do not know what does in fact qualify as a trade secret under the law, but the information itself is known to the employees who use it in the company's business operations.

There are three steps, then, to generating an inventory of the company's trade secrets. The first is to provide an adequate level of training to employees on what a trade secret is, so that they can distinguish trade secret assets from other types of information. The second step is to collect a list of potential trade secret information assets from the employees. The third step is to reconcile these lists and remove the numerous redundancies that will result from multiple employee lists.

This third step can be a daunting task. For example, is there any doubt that most employees of The Coca-Cola Company would identify the formula for Coca-Cola as a company trade secret? How many employees of KFC Corporation would identify Colonel Sanders' secret recipe and its blend of 11 herbs and spices as a company trade secret? These are easier cases;

the difficult ones will be more subtle. Is the trade secret information contained in Engineering Lab Test #11 the same proprietary information that is used in the current manufacturing production equipment, or is it a different formulation — an earlier obsolete one, the prototype for the next generation, or simply a failed experiment?

Some of this difficulty can be avoided in the collection process. Employees should be asked to identify only those trade secrets that they created or use in their jobs, or which are in their immediate control. Further reduction in the redundancy can be achieved by having supervisors submit the trade secret lists for their groups, capturing the information from employees either individually or in group meetings. Each supervisor reports only those trade secrets created or used within, or in the immediate control of, their group. These lists can in turn be submitted up the management hierarchy, with each level resolving redundancies within their own reporting structure before passing on their organization's list. These methods create some risk that certain trade secrets will not appear on the company's inventory, but it is unlikely that the important and valuable trade secrets will fail to appear in the consolidated inventory. The lists at each level should be transmitted to a central repository for archival purposes. The lists can also provide the foundation and starting point for the next trade secret audit, with additions and subtractions made to the master list.

An alternative to creating a trade secret inventory directly from employees is the use of outside professional consultants to inventory the company's trade secrets. The outside inventory will consist of interviews of employees in key locations throughout the company, and collection and collation of the interview data by the consultants. One disadvantage of the outside trade secret inventory approach is that the outside consultants do not know the company's business as well as the employees, as well as what is already known and not well known in the trade. On the other hand, an advantage of outside professional consultants is that no special trade secret training of employees is necessary. Finally, for trade secret audits conducted during litigation, speed is an overriding concern and outside consultants are often necessary in these circumstances.

The problem with both of these inventory methods is that a trade secret portfolio is dynamic. Such manually-prepared inventories go out-of-date quickly and need to be periodically updated. These manual inventories are snapshots — or more correctly hand-painted portraits — of the trade secret portfolio of the company at a single point in time.

Technology is beginning to assist with the inventory process. Processes that collect trade secret inventory information directly from employees into trade secret databases via the company intranet are becoming available. They will make it less likely that trade secrets are overlooked or omitted in manually consolidated lists, and eliminate the handwork by employees or outside lawyers and consultants that is required to consolidate the information. A further advantage of such technological solutions is that they allow the collection and consolidation of the trade secret inventory data to be performed continuously, resulting in a continuously updated inventory of the trade secret portfolio. Rather than a snapshot, these products will provide a "motion picture" of the trade secret portfolio that can be viewed in real time or "rewound" to an earlier date of interest.

Categorization of Potential Trade Secrets

Once an inventory of potential trade secrets is in hand, it is necessary to organize it into categories for efficient use. This is similar to the organization of items in a physical inventory, which may include the categories of Real Estate, Vehicles, Furniture, Manufacturing Equipment, Office Equipment, and the like. With trade secrets, the design of a system of categories is complicated by the tremendous variety of information that may qualify for trade secret protection within a single company, as well as by the differences in this information from company to company. The trade secret checklist for a typical manufacturing company in Appendix C illustrates the diversity of trade secret information.

The authors have determined that trade secret assets most efficiently fall into a categorization system based on "<Subject><Format> for <Product>", such as Manufacturing Process for Disk Drives, Marketing Business Case for Cola Beverages, Sales Forecast for Lawn Furniture, or Engineering Specification for Transmission. These categories of trade secrets are called SFPs. All of the possible SFPs form a three-dimensional trade secret space of the company, into which all of the trade secrets fit.

The Subject usually corresponds to the department or organization within the company that created or uses the trade secret. The Format includes document formats as well as prototypes, processes, formulas, results, plans, and other categories that are appropriate to the company. The Product most often refers to brand name, but trade secrets may also apply to groups of products, such as a trade secret that applies to the bottling method for all carbonated beverages produced by the firm and not to one specific beverage.

The SFP categorization method provides fine categorization of large numbers of trade secrets with a small number of Subjects, Formats, and Products. For example, a company that has 10 departments, for which 30 different information formats are meaningful, and which produces 20 different products, has a total of 6,000 SFPs available as trade secret categories, into which tens or hundreds of thousands of trade secrets can be efficiently sorted. An additional benefit is that employees are already familiar with the departments within the company, the different types of information formats used within the company, and the products produced by the company. No employee training is required to use these 6,000 categories since everyone already knows what constitutes an Advertising Plan for Snack Products or a Packaging Design for Laundry Detergent.

The categorization of the company's trade secret inventory into SFPs can also be accomplished during the inventory collection process. The department in which the trade secret is created or used, or which has immediate control of the trade secret, is likely to be the proper Subject, while the proper Format and Product are known to the employee submitting the trade secret to the inventory. If the SFPs are collected during the trade secret inventory, redundancy elimination needs to be performed only within each SFP, simplifying the collation process as well.

Identification of Actual Trade Secrets

With the inventory and categorization processes completed, the company is positioned to evaluate the potential trade secret assets. No effort has been made to this point to see if each of the proposed trade secrets meets the legal criteria for trade secret protection.

The best guide to the determination of a trade secret asset for accounting purposes is the existence test that must be proved in a legal proceeding for trade secret misappropriation as previously discussed in Chapter 3. In order to be accounted as an asset, the potential trade secret must be defensible as company property. In applying the existence test, the courts rely on the legal tests in the Uniform Trade Secret Act (UTSA) as well as the six factor test in the Restatement (First) of Torts.

The definition of a trade secret in the UTSA is repeated here for convenience.

1) "Trade secret means information ... that ... derives independent economic value, actual or potential"

2) "from not being generally known to [other persons who can obtain economic value from its disclosure or use]"

3) "and not being readily ascertainable by proper means by other persons who can obtain economic value from its disclosure or use, and"

4) "is the subject of efforts that are reasonable under the circumstances to maintain its secrecy."

For each trade secret in inventory, then, we have four questions to ask to determine whether each potential trade secret qualifies as a trade secret in fact:

1) Does the company derive independent economic value from the information being a secret from its competitors?

2) Is the information not generally known in the trade — i.e., to competitors?

3) Is the information not readily ascertainable by proper means?

4) Is the information the subject of efforts that are reasonable under the circumstances to maintain its secrecy?

The UTSA definition provides a basic "go/no-go" test for

trade secret existence, but it will not evaluate gradations. A more thorough examination can be performed by considering the six factors of a trade secret. They are repeated here for reference in Table 11-1.

Assessing the six factors provides a more accurate and sensitive evaluation of the potential trade secrets in the inventory, but it is a more difficult examination process. This evaluation is necessary for the company's most valuable trade secrets, to determine how well they are likely to stand up in court, and whether the documentation and security of the trade secret information is adequate.

Like the categorization process, the identification process can be simplified. The simple method of one-to-five scoring on each of the six factors is surprisingly effective at capturing the extent to which the trade secret meets the six-factor test. It also has the benefit of being well understood by employees already. We are all familiar with one-to-five scoring from multiple life experiences, from hotel and movie rankings to the ubiquitous A-B-C-D-F grading method used in schools. Once again, technology is being developed to assist in the identification process, and automated methods of scoring the existence and strength of trade secrets using one-to-five scoring of the six factors are now becoming available.

Table 11-1 Six Factors of a Trade Secret from the Restatement (First) of Torts

"An exact definition of a trade secret is not possible. Some factors to be considered in determining whether given information is one's trade secret are:

(1) the extent to which the information is known outside of his business;

(2) the extent to which it is known by employees and others involved in his business;

(3) the extent of measures taken by him to guard the secrecy of the information;

(4) the value of the information to him and to his competitors;

(5) the amount of effort or money expended by him in developing the information;

(6) the ease or difficulty with which the information could be properly acquired or duplicated by others."

It should be noted that identification in the dynamic trade secret portfolio will not be static. A trivial case is that of the quarterly financial report, which moves from being an extremely sensitive trade secret prior to release to being public knowledge after release. Similarly, a new trade secret packaging design becomes public knowledge as soon as the new packaging design appears in ads and on store shelves. In addition, the security procedures applied to a particular trade secret may be increased as its importance becomes apparent, or may be decreased as the information moves from development to manufacturing.

Identification of the existence of the trade secret asset is the critical step in the accounting process for trade secrets. Absent an evaluation of the trade secret status of the potential trade secrets, the inventory is merely a list of employee suggestions. It is the identification process that separates legally protectable trade secrets from general knowledge that can be legally copied, disclosed, or used by anyone for any legal purpose.

The Benefit of Employee Categorization and Identification

Information about trade secrets comes from the employees that create or use the trade secrets in their jobs. Pushing the categorization and identification of trade secrets down to the employees actually in contact with the trade secret has substantial benefits.

The obvious benefit is cost. Sending out in-house or outside interviewers to employee locations to conduct extensive interviews is a costly business. Assuming the categorization and identification process can be made simple enough, the employee instead can enter the information themselves. This likely takes less employee time than the interview process would, and the costs of trained interviewers is saved.

More importantly, though, is security. The trade secret information is secure in place. Exposing the trade secret to an interviewer, whether in-house or from outside the company, increases its exposure, and consequently the risk of loss. The interview process trains the interviewers in all the company's trade secrets. Given what we have already seen of the risks of insider loss in Chapter 7, this is a chilling prospect. Trade secret accounting should make trade secrets more secure, not increase the risk of loss through increased exposure.

This is where the importance of SFP categorization and one-to-five identification lie. By making the categorization and identification process simple enough, they allow the process to be pushed down to the employees already in contact with the trade secret, increasing accuracy, reducing cost, and maintaining the security of the trade secret.

Classification of Trade Secrets

Different trade secrets require different handling, but to this point the accounting process has generated an inventory of trade secrets that are all of a kind. The UTSA requires measures "reasonable under the circumstances", and the courts have logically found that the relative value of a trade secret to the company is in fact one of the circumstances that determines the proper level of security required. The structure of the company's sales incentive program, the compensation package for a key executive, and the results of a failed experiment are all trade secrets, but they are not of the same sensitivity as the recipe for the company's flagship product, its new product introduction plan, or its pre-release quarterly results. Trade secrets must be sorted into classifications that indicate their sensitivity and guide their handling and security.

It is recommended that a classification scheme have no fewer than three and no more than five levels, including the classifications of "Not Confidential" for general information that has no protections and "Personal Information" for items such as employee or client addresses, home telephone numbers, social security numbers, and health information that the company has a legal obligation to protect.

This leaves one to three levels of classification that specifically apply to trade secret assets. Different companies use different labels for these classifications, but the most frequent for a single level is "Confidential and Proprietary", while for three classes of trade secrets, the levels "Confidential", "Secret", and "Top Secret" are commonly used. For each of these levels, there should be a structured regime of security measures and rules for distribution, disclosure, transportation, and access control and tracking that are tailored to the sensitivity of trade secrets of that classification.

Criteria for classification of potential trade secrets can also be collected from employees during the inventory process and reviewed by a skilled evaluator during the final review process. Once applied, these classifications should be used to direct the management of the trade secret asset according to the proper regimen for the classification level. In particular, once classifications are applied, it is necessary to ensure that the proper regime be instituted for the trade secret information if it is not currently in place, and that all trade secret information be properly marked with its classification on any media in which it appears. Some documents for more general distribution may need to be redacted to remove more restricted information contained in the document. Once again, technology is appearing to simplify this task, with intelligent redaction software being developed to address these tasks.

As with identification, classification will not be static in the dynamic trade secret environment. The results of failed experiments will become much more sensitive if a successful experiment provides an important breakthrough, while an extremely sensitive product formulation will become less sensitive if the product is unsuccessful or becomes obsolete in the marketplace. Classifications should be reviewed periodically to ensure that the classification, and the handling and security procedures that are driven by it, remain appropriate to the sensitivity of the trade secret information.

Having completed inventory, categorization, identification, and classification, the company is in possession of a complete and structured list of the actual trade secret assets that are properly classified to ensure appropriate handling and security protocols.

Summation

Creation of a trade secret inventory necessary to any proper accounting requires the steps of inventorying potential trade secrets, categorizing the potential trade secrets, identifying the actual trade secrets among the potential trade secrets, and classifying the identified trade secrets to ensure proper handling and security protocols. The inventory step can be performed through direct employee reporting, employee interviews, or collection from employees through automated means. Categorizing trade secrets into SFPs provides fine granularity in an intuitive system that does not require employee training. Identification is the critical step of separating trade secret information from information that does not meet the legal criteria for trade secret protection. Finally, classification provides a structure within the company that can ensure that trade secret information is appropriately handled and secured as required by its sensitivity. The end result of this process is a complete, useful, and organized listing of the company's trade secret assets, separated into proper classifications for appropriate handling and security protocols.

Chapter 13
Valuation and Reporting

Executive Summary

Valuation of trade secrets is a critical step in realizing their true value to the company. As the definition of trade secrets anticipates future cash flows, the recommended valuation method is the discounted cash flows to be derived from the trade secret information. FASB rules require valuation of acquired trade secrets, and application of the FASB rules to internally created trade secrets as well creates a single regime for trade secret valuation within the company. Reporting of trade secret valuations can be done in such a way as to provide additional shareholder and investor guidance without disclosure of the trade secret information to competitors. Finally, assetization of trade secrets permits their treatment in the same manner as the physical assets of the company, including insuring the assets, collateralizing them, and licensing them to other companies.

Importance of Valuation

Valuation is necessary in accounting for any assets, including trade secret assets. Valuation of trade secret assets is also necessary to realize the full value of a trade secret inventory. The assetization of trade secrets requires that the value of the trade secret assets be accurately determined. Once assetized, trade secrets can be insured for their value to the firm, licensed to other companies, and used for collateral in loans. The trade secret valuation will also be a major factor in setting the value of the company in any merger and acquisition negotiations.

Valuation also plays a critical role in trade secret asset security measures. The decision as to which security measures are appropriate for a given trade secret asset should be driven by the value of the trade secret asset. Security measures "reasonable under the circumstances" are required under the Uniform Trade Secret Act (UTSA), and the courts consider the value of the trade secret information as one of the circumstances to be taken into account in determining what security measures are reasonable. Further, if the trade secret asset becomes the subject of a trade secret misappropriation lawsuit, its valuation will drive the consideration of compensatory and punitive damages.

Finally, reporting of trade secret assets is impossible without valuation. What is there to report if no value has been determined? With no reporting of trade secret assets, the company's stock valuation is lacking the concrete basis that can reduce perceived shareholder risk and share price volatility.

Difficulty of Valuation

Valuation of trade secret assets is more difficult than the valuation of physical assets. Every physical asset has an historical acquisition cost that can serve as a basis for valuation. Further, physical assets usually have a replacement cost that can be accurately determined in the marketplace. There may also be a used market for physical assets to assist in determining valuation.

Trade secrets are, first and last, information. There is no

physical item to inventory. The information itself is not as well-bounded or well-defined as a physical asset. Trade secrets are by definition unique to the company, so there is no replacement cost that can be determined, no used market to assist in setting a valuation. Finally, there may be no way to determine the historical acquisition cost. For a trade secret that was developed in the research department, what is the true acquisition cost: the minute cost of a moment's flash of insight, or the large research budget of years of trial and error?

Finally, trade secret asset valuation is now complicated by regulation. The book value of a physical asset reported in the company's accounting statements is generally the depreciated acquisition cost. This may have little relationship to the actual replacement cost or the fair market value of the item in the used equipment market. So, too, with trade secrets. FASB Nos. 141 and 142 address the proper accounting of trade secret assets acquired in a merger or acquisition, and the resulting valuations may have little relation to the actual value of the trade secret to the company. Reporting of trade secrets created within the company is not covered by FASB Nos. 141 and 142, however, resulting in two classes of trade secrets for accounting purposes.

Valuation Method for Internal Trade Secrets

With no historical acquisition cost, replacement cost, or value in the marketplace, many of the valuation methods used for physical assets cannot be applied to trade secret assets. The definition of a trade secret in the UTSA addresses valuation, however: "Trade secret means information ... that ... derives independent economic value, actual or potential, from not being generally known to ... other persons who can obtain economic value from its disclosure or use...." The Restatement (First) of Torts is more explicit: "A trade secret may consist of any formula, pattern, device or compilation of information which is used in one's business and which gives him an opportunity to obtain an advantage over competitors who do not know or use it." The Restatement (Third) of Unfair Competition is even stronger: "A trade secret is any information that can be used in the operation of a business or other enterprise that is sufficiently valuable and secret to afford an actual or potential economic advantage over others." The "independent economic value", "advantage over competitors", and "economic advantage over others" of these legal definitions require a trade secret to provide the opportunity to realize future cash flows ("economic advantage") from the information. The proper method for determining the current value of a trade secret asset, then, is by the net present value of those future cash flows, also called the discounted cash flows method.

It must be noted, however, that all three of these legal definitions also require security measures. The UTSA definition requires measures "reasonable under the circumstances to maintain its secrecy." The Restatement (First) of Torts includes the requirement that the information provide an "advantage over competitors who do not know or use it", and includes "the extent of measures taken by him to guard the secrecy of the information" as one of the six factors of a trade secret. The Restatement (Third) of Unfair Competition requires the information to be "sufficiently valuable and secret" to qualify as a trade secret. The courts will not protect information after the fact that the information owner has failed to protect in the first instance. Absent reasonable security measures, there is no trade secret protection of the information, and the information can be legally used by anyone. The economic value of the alleged trade secret asset in this case is exactly zero.

Assuming that reasonable security measures are in place, analytical methods can be used to determine the value of the trade secret asset. In specific cases, the professional assistance of valuation experts is usually required to perform these valuations. Nevertheless, we can make some general observations about valuations that apply to all companies.

First, for public companies, the total valuation of intangible assets usually should not exceed the difference between the book value of the company — the accounting of its physical and financial assets — and its stock market capitalization. The stock market has already made an assessment of the fair market value of the company, and the value of its intangible assets, including trade secret assets, should not cause its total valuation to exceed the market's judgment. In the case of Google, discussed in Chapter 2, the total value of its intangible assets as of the end of Q2 2015 cannot exceed $240 billion. While it is fair to argue that the markets have underestimated the future earnings potential of a particular company, absent compelling analysis, the aggregated risk assessment of investors who are purchasing the company's shares in an open market should take priority over management assertions.

Second, the use of SFPs to categorize trade secret assets, discussed in the previous chapter, provides additional insight into valuations. The combined valuation of all of the trade secrets contained within a specific Product category should be determined from the discounted cash flows to be realized from that product alone. The valuation of trade secret assets that apply to multiple products should be determined pro-rata from the discounted cash flows from all products in the category.

As an example, the value of the recipe for Coca-Cola should be determined from the discounted cash flows anticipated from the sale of Coca-Cola. In addition, however, the value of other trade secret assets associated with Coca-Cola — such as the marketing plans, the new advertising campaign, the bottling methods, the names and addresses of suppliers and resellers — come under the Coca-Cola Product category. Finally, a pro-rata share of the value of trade secret assets applying to Carbonated Beverages and similar group categories also come under the Coca-Cola Product category. The total value of all these trade secret assets cannot exceed the discounted cash flows anticipated from sales of Coke.

Similarly for divisions within the company, the total value of the trade secret assets within each division cannot exceed the discounted cash flows anticipated for products and services produced by that division.

In specific exceptional cases, the discounted cash flows method may not be appropriate. For example, a product may fail in the marketplace, rendering the previously expected future cash flows valueless. The trade secrets associated with the product may nevertheless have value in application to other products, or through sale or license to another company. Another example is a new trade secret in the research department, which may or may not be applied to a new product immediately, or which may be applied to a new product whose future cash flows are unpredictable.

Valuation Method for Acquired Trade Secrets

Since 2002, valuation for trade secrets and other intangible assets acquired in a merger or acquisition has been covered under FASB Nos. 141 and 142. Previously carried in the aggregate on the books as Goodwill and amortized over forty years, some acquired intangible assets must now be accounted for individually, including a determination of the fair value and useful life of the asset. These valuation determinations must be conducted or reviewed by a valuation expert, independent of the company or its auditors. FASB Nos. 141 and 142 offer flexibility in how fair market values are determined, but express a preference for valuation using discounted cash flows as in the preceding discussion for internal trade secrets,

Further, the useful life of intangible assets covered by FASB Nos. 141 and 142 is no longer restricted to forty years. In particular, indefinite lifetimes are possible. FASB Nos. 141 and 142 do require an impairment test annually to determine if the fair market value has fallen below the book value of the asset, requiring a write-down of the difference.

Finally, FASB Nos. 141 and 142 require that the valuation of acquired intangible assets be assigned to the individual reporting units within the acquired company. Such assignment is consistent with the use of the Product category of the SFPs for valuation of internal trade secrets in the preceding discussion.

Implementation of FASB Nos. 141 and 142 is well beyond the scope of this book, but it is important to note that the FASB rules now require accounting for some acquired intangible assets, including trade secret assets. Further, the methods specified in FASB Nos. 141 and 142 for accounting for these acquired intangible assets are consistent with recommended practices for the accounting of internal trade secret assets. This allows a single accounting method to be applied across all trade secret assets of the company, however acquired. Additionally, accounting for trade secrets by these methods prepares the way for merger or acquisition of the company in the future by allocating valuations across reporting units in advance of the requirements.

Reporting

Two questions naturally arise in connection with reporting of the valuation of trade secret assets. First, will reporting disclose the nature of the company's trade secrets, thereby alerting competitors and eroding the secrecy of the information? Second, why should the company report the value of its trade secret assets absent a regulatory requirement to do so?

When reporting the value of trade secret assets, it is important to report them in large enough aggregate categories and with general enough descriptive labels to obscure the nature of the information from competitors. "Methods of Determining Consumer Taste Through Focus Groups", for example, is too descriptive, and discloses the company's use of focus groups; such trade secrets should be subsumed into a category more generally titled, such as "Market Research Methods". Similarly, "Methods of Annealing Structural Metal Products" is too specific, and discloses annealing as a process used by the company; such trade secrets should be reported within "Structural Metal Production Methods" instead. Such general descriptions provide shareholders and investors with information that aids in an understanding of the nature and value of the company's trade secret assets — and the company itself — without disclosing the nature of the information.

In specific companies, even such categories may disclose too much. For example, for some companies, the relative values of marketing, research, and manufacturing trade secrets may provide valuable information to competitors. In such cases, the valuation of trade secret assets can be reported on a product by product basis, such as "Trade Secrets Relating to Structural Products" and "Trade Secrets Relating to Decorative Products".

In other cases, categorization by product may disclose too much, and valuation of trade secret assets can be reported on a functional basis, such as "Marketing Trade Secrets" and "Manufacturing Trade Secrets". It is important to choose a reporting structure that provides a granularity of reporting that will be of value to shareholders and investors without being of value to competitors.

As to reporting trade secret asset valuations, the company benefits are twofold. First, for a company whose market capitalization greatly exceeds its book value, reporting of trade secret asset valuations addresses the shareholders' perceived risk of investing in the company. If investors have access to a considered analysis of the value of the company's intellectual property, the company's share value is given a more substantive basis that can reduce share price volatility.

Second, the valuation of trade secret assets is a necessary step in assetizing the trade secret information of the company. Assetizing trade secrets allows the company to treat these assets in the same manner as the other assets of the firm, including insuring, borrowing against, licensing, and selling the assets to other companies.

Assetizing Trade Secrets

While people loosely talk about intangible assets, intellectual property assets, and trade secret assets, in fact they seldom actually treat these items as assets in the manner that the other assets of the firm are treated. Where failure to insure a company building or vehicle against loss would border on criminal negligence on the part of management, trade secrets are seldom if ever insured. The company's physical assets are often used to secure loans, and such assets are routinely bought, sold, or leased in the normal course of business, but none of these asset-based transactions occur with trade secrets.

Several major insurance companies have attempted to offer trade secret insurance without success. The technical problem has been underwriting, specifically the determination of value and the actuarial appraisal of risk. As a result, companies are operating without insurance protection for the most valuable assets they own, insurance protection that they would not even consider doing without on their less valuable physical assets.

In addition to spreading risk of loss across companies and reporting periods, insurance companies routinely perform risk assessments on their insured's operations. Fire safety precautions are reviewed in the context of building insurance, and occupational safety and health evaluations are performed in the case of workmen's compensation and health insurance. Programs to reduce risk, including employee education, workplace materials, business process reviews, and equipment evaluations are offered by insurers to their insureds to manage risk. Insurance companies thus aggregate industry experience with risk and provide the benefits of this experience to their individual insureds.

None of these benefits are available for trade secret assets absent insurance protection. The company is left naked to huge losses from a single incident, without any spreading of the risk across companies and reporting periods. No aggregation of industry experience in managing the risk of trade secret loss is available to the company, leaving each company to go it alone in developing and assessing its own trade secret protection program. Programs to reduce risk of loss of trade secret assets are not standardized, and employees cannot carry their experience with trade secret protection from one company to another due to the differences in programs.

Assetizing trade secrets will allow trade secrets to be insured against loss. The inventory and valuation of the trade secret assets of the firm, together with effective security and management to reduce the risk of loss, turns trade secret assets into true assets of the company and makes insuring trade secret assets possible.

Once protected, valued, and insured, the trade secret assets of the firm are also available as collateral for loans. The firm need no longer be restricted to the physical assets represented by its book value or the dilution of shares in securing funding.

Finally, there is no reason why the trade secret assets of the firm should not be available for sale or license to other companies. While the company may not wish to license trade secrets to direct competitors, many trade secrets have application outside of their particular use within the company. An additional source of revenue from the licensing or sale of trade secrets to companies in adjacent markets is available once the company has properly inventoried and categorized its trade secrets. The valuation of the trade secrets also gives valuable guidance in setting the royalty rates or sales price for the trade secrets.

Special Case: Transfer Pricing

A special case in valuation and reporting of trade secrets arises in the context of transfer pricing. If the company is international in scope, and trade secrets are transferred across international boundaries, tax laws in the two countries regulating the transfer pricing of assets apply to intellectual property, including trade secrets. In this context, the accurate valuation and reporting of trade secret assets transferred across international borders becomes critical in ensuring tax compliance and avoiding the risk of tax liability.

Summation

The valuation of trade secret assets is an important part of intellectual asset management. Valuation can be performed in many ways, but is most often calculated from the discounted cash flows to be derived from the trade secret. FASB rules now require valuation of acquired trade secrets, and these regulatory requirements are consistent with best practices for valuation of internally created trade secrets. Reporting the value of trade secrets, if done properly, provides additional guidance to shareholders and investors without putting the trade secret information at risk. Finally, assetizing the trade secrets provides the benefits associated with the more traditional physical assets of the firm, including insurability, collateralization, and realization of additional revenue from sale and licensing of trade secret information.

Chapter 14
Life-Cycle Management of Trade Secrets

Executive Summary

Trade secrets are dynamic, and every trade secret has a life cycle, from creation, through development, patent election, application, and potential licensing, to obsolescence. The proper handling of trade secrets should take into account where the trade secret is in its life cycle. The classification of a trade secret will also change during its life cycle, until it is finally declassified as obsolete. Life cycle management of trade secrets maximizes the economic benefit the company receives from the trade secret, enables effective security, and lessens costs by reducing unnecessary security efforts.

Creation

When a trade secret is created, its ultimate importance to the company may not yet be known or understood. A trial formulation for a new soft drink, for example, may or may not ultimately be chosen as the formulation to be marketed, and even if it is the chosen formulation, the new soft drink may fail in the marketplace or succeed wildly. At the same time, trade secrets, once disclosed, cannot be "re-secreted". For this reason, trade secrets when created should be classified at the highest classification level which they may ultimately deserve.

Consider the trial formulation for a new soft drink in a company with three trade secret classification levels,

"Confidential", "Secret", and "Top Secret". The company's product formulas may all be classified as either "Top Secret" for its signature products or "Secret" for its more generic products, per company policy. In addition, failed or rejected product formulations may be classified as "Confidential". In this scenario, the trial formulation for the new soft drink should be classified as "Top Secret", because it is not yet known whether this formulation will ultimately be rejected, be marketed as a generic product, or be marketed as one of the company's signature products. If the formulation's status changes from trial formulation to rejected or generic, its trade secret classification can be downgraded, if necessary, to the appropriate level.

This is the logical explanation for the common gut feeling that research and development activities are the most sensitive areas in the company for trade secret protection. Trade secret information in the research and development areas is still unknown as to its importance, and so all trade secret information in research and development must be handled at the highest level of confidentiality.

Development

As trade secrets are developed, new trade secrets are created in support of and to facilitate the use of the original trade secret. In our example of the trial formulation of a new soft drink, a new packaging design may be created, a marketing campaign may be planned, print and television advertisements may be developed, new bottling equipment may be needed, new distribution arrangements tailored to the new drink's market may be required. An effort should be made to keep this growing package of trade secrets internally consistent in their handling. For example, it would probably not be appropriate for the marketing campaign to have a higher classification level than the formulation itself. Further, the new bottling equipment may incorporate significant portions of the trial formulation information into their very design, and so should be classified at the same level as the new formulation itself.

As the development of the trade secret continues, its importance to the company may be better understood. Market research on the new formulation may indicate that the new soft drink will be a major seller for the company, generating pull for the company's other products into new markets, and deserve the status of being a company signature product, retaining its "Top Secret" classification. Conversely, market research may make it clear that the product will be a solid seller in combination with the company's other products, relying on their marketing coat tails, and thus be more of a generic product, justifying a downgrade of its classification status within the company's classification strategy to "Secret". Finally, it may become clear that the product will not succeed in the marketplace, and the project may be dropped entirely, resulting in the downgrading of the formulation's classification level to "Confidential". For our example company, the status would never be lowered to "Not Confidential" or "Unclassified"; all drink formulations are considered trade secrets, if for no other reason than to deny the company's competitors any information about the direction of the company's research and development efforts.

Patent Election

Every patent begins life as a trade secret. Conversely, inherent in the life cycle of many trade secrets is an evaluation of whether a trade secret constitutes patentable subject matter. In order to be patentable, the trade secret information must also be novel, useful, and non-obvious. These determinations often require a review of patentability by the company's patent counsel.

If the trade secret information is patentable, it still may not be in the company's best interests to convert the trade secret asset to a patent asset. When a patent issues, any trade secret rights in the information disclosed in the patent application are lost. If you elect to pursue patent protection in a foreign country, the U.S. Patent and Trademark Office will publish the patent application after eighteen months, which is generally before any final decision on allowance of the patent. The result may be a lack of trade secret or patent protection for the invention, since the trade secret rights are lost on publication of the application and patent rights never come into existence if the patent application is disallowed by the patent office.

Application

Application of the trade secret in the manufacturing of a product requires wider distribution within the company. Employees and vendors involved in the manufacturing of the products, purchasing of the components or ingredients, packaging of the finished product, writing and printing of the manuals and packaging, and a host of other tasks associated with a new product introduction will all need access at some level to some of the trade secret information in order to get the product out the door. The importance of proper classification of trade secrets associated with the product is now clear. Care should be taken at this point not to disclose too much too widely for trade secrets in the upper classifications.

In our soft drink example, clearly the most important and sensitive trade secret is the product formulation, including the purchase of the necessary ingredients and the incorporation of the product formulation into the bottling equipment. Various methods may be used to obscure these trade secrets, including aggregating the ingredients purchase with purchases for other products, splitting orders for each ingredient among multiple suppliers so no supplier knows how much of any ingredient is used, or even buying excess amounts or additional ingredients and throwing away the excess, and producing one or more intermediate stages of the product, such as the syrup, at a central, secure location. All of these methods apply to other products as well, including mechanical and electrical products.

Such extreme — and costly — measures need not be taken to protect the lesser trade secrets associated with a product. The advertising campaign will soon become public knowledge and be declassified in any case, as will the product packaging, the manuals, and other associated printed materials. The key is to differentiate the trade secrets effectively, so that security efforts, especially access controls, can be tailored appropriately to the sensitivity of the individual trade secrets.

Licensing

While the company has trade secrets that are so sensitive — such as our example soft drink formulation — that licensing will rarely be appropriate, other trade secret information in the possession of the company may be candidates for licensing or sale to other companies. The company's expertise in packaging soft drinks may be valuable to other companies that compete in other product areas or other geographical areas. Licensing of trade secrets in these cases provides another economic benefit of the trade secret information to the company.

Obviously, care must be taken during the negotiations, as well as with the selection of potential licensees. The trade secret information cannot be disclosed prior to the completion of negotiations. This is the converse of the inbound security issue with trade secret licensing negotiations discussed in Chapter 8. The value of the trade secret information to the licensee should be investigated by an independent third-party to whom the information can be disclosed and who can advise the licensee without disclosing the information.

Care must also be taken in selecting potential licensees. The obvious concern is to ensure that the licensee is not now, and is unlikely to become, a direct competitor of the company. Even licensing trade secret information in a different product area than that in which the companies compete improves the competitor's financial position, providing additional financing for product areas where the companies do compete head-to-head.

Another serious concern relates to the security practices of the potential licensee. Assurances by the licensee that the licensor's proprietary information will be protected carry little weight if the licensee has an ineffective trade secret protection program in place. An evaluation of the potential licensee's trade secret protection program and protocols should thus be a part of any licensing considerations.

Obsolescence

Because of the possibility of independent development, trade secrets can become obsolete at any time. Obsolescence in this context does not mean that the information is no longer useful; obsolescence here refers to the lack of independent value derived from the secrecy of the information because the information is no longer secret. Once the trade secret becomes generally known in the trade, the trade secret rights in such information are extinguished as a matter of law.

However, it may still benefit the company to keep its actual use of the information confidential. In the case of our soft drink example, assume that the beverage contains a particular ingredient not usual to soft drinks, aggregated into "Natural and Artificial Flavors" on the mandated product ingredients listed on the packaging. Even if it is publicly disclosed that this ingredient may be appropriate for soft drinks, even for soft drinks of the specific type as our hypothetical product, that does not disclose that this soft drink company is actually using this ingredient in any particular product. Maintaining the secrecy of this ingredient's use can therefore still qualify the usage information as a trade secret. In other cases, the disclosure of a specific method or ingredient in the public domain will destroy the benefits of maintaining any aspect of secrecy relating to the information. In this latter case, the information should be declassified as a trade secret.

Another case of obsolescence occurs when the information no longer has any actual or potential economic value. It is still appropriate to maintain the secrecy of the information if the information has actual or potential economic value to the company's competitors, but no security is necessary if the information is stale or devoid of economic value. For example, the proprietary methodology for compliance with certain regulatory or technical standards may become useless when the standards are superseded. In this case, declassifying the information may be appropriate.

Finally, trade secret information can become obsolete when it is intentionally disclosed by the company itself. An example here is the company's quarterly financial report, which is highly sensitive secret information prior to release, but is published in the company's regulatory filings and on the company's web site on the release date. Declassifying such disclosed trade secrets is appropriate upon their release. Even though the information continues to be useful, the trade secret status of such information has been forfeited by the public disclosure of the information.

Summation

Trade secrets are dynamic and have a life cycle. Proper handling of individual trade secrets at each point in their life cycle is important if security is to be maintained and the maximum benefit is to be derived from the trade secret. The life cycle of a trade secret includes creation, development, patent election, application, potential licensing, and obsolescence. Appropriate classification of each trade secret throughout its life cycle will ensure the proper handling of the trade secret and the maximum benefit at the least cost of security.

Chapter 15
The Trade Secret Audit

Executive Summary

In order to know that a company's trade secret measures are complete and operational, periodic trade secret audits must be performed. Much like an accounting audit, a trade secret audit makes sure that all the pieces are in place, and are being correctly executed. Policies, procedures, documentation, inventory, and valuation must all be reviewed to make sure the company's trade secret protection program is effective, and not mere pageantry.

The Need For Trade Secret Audits

One role of the trade secret audit is to make sure all the components of a complete trade secret protection program are in place. Like a jigsaw puzzle, it is only when all the pieces are assembled in one place that you can tell whether all the pieces are there. The first part of the trade secret audit should be to pull all the pieces together and review the program for completeness.

Having the right policies and procedures in place isn't enough, however. Business processes, like mechanical processes, are best regulated using feedback, not open-loop control. Policies and procedures can be put in place, but without checking on their implementation, they can easily be ignored or forgotten in the press of day-to-day business. Given the exposure to trade secret theft of the company's competitive position and its resulting profits, neglecting to carry out the trade secret program can quickly become an existential problem.

Finally, trade secret protection is just one side of a game of Spy vs. Spy. As misappropriators enhance their methods and techniques, the company, too, must up its game. Effective trade secret protection is a moving target in an environment in which those who would steal technology are evolving their methods and tools. The audit should seek to identify areas in which the company's programs and procedures were adequate in the past, but need to be augmented going forward.

Why You Need Outside Help

Much like an accounting audit, a trade secret audit is not something you can perform internally, and for much the same reasons. Outside help brings the much-needed benefit of an outsider's view to your audit, as well as providing information about what the rest of the business community is doing that is unlikely to be an inside expertise. The outside firm should be aware of the current state of threats, new developments in the threat landscape, and recommended best practices to counter emerging threats.

Bringing in an outside audit team also keeps you and your staff from fooling yourself that the program is doing its job, when in reality you and your company are terribly exposed due to gaps in the structure or execution of your program.

The outside audit team should be an interdisciplinary team that includes legal, technical, and business expertise, all of whom specialize in trade secrets protection. They should have a plan for the audit that they can provide during the selection process, and which they include as part of their written proposal.

Deliverables specified in the proposal should include a line-by-line report card on the elements of the trade secret program, an evaluation of any deficiencies found, detailed recommendations for correcting the deficiencies, and a forward-looking view of developing threats and possible countermeasures.

The Role Of The Attorney

An attorney must be involved in the selection of the outside trade secret audit team, the conduct of the audit, and the receipt of deliverables from the outside team. Best is if counsel, whether inside counsel or outside counsel, is in the chain of supervision of the entire process. If counsel is directly involved, the trade secret audit process falls under attorney-client and attorney work-product privileges, and the processes, materials, and results from the trade secret audit cannot be subpoenaed without a court order and an opportunity to quash the subpoena on privilege grounds.

It only makes sense. Trade secret is itself a legal construct. A trade secret audit should be conducted under proper legal supervision.

Gather The Documentation

Appendix D provides a preliminary list of the documents you need to pull together so the outside team can assess the baseline for the trade secret audit. Gathering these pieces together internally prior to the arrival of the audit team means they don't have to do it piecemeal once they get there. It will save the company money to use internal staff to do this collection versus having the outside team of experts spend time digging for it.

The Importance Of The Trade Secret Inventory

One of the items listed in Appendix D is the existing trade secret inventory. It is important to note that this inventory should include only metadata — information about the trade secrets — and not the trade secret information itself. There are several reasons for doing this.

One is that you do not want to create, in building the trade secret inventory, a prime target for trade secret thieves. As a trade secret misappropriator, why track down trade secrets from all over the company, if you can find them all in one place, the trade secret inventory? So don't create that huge target.

The second is that, to build such an inventory of the trade secret information itself, you have to have people within the company go around and collect the trade secret information, exposing the information to more internal people. The extent to which the trade secret is known within the company is one of the six factors of a trade secret, so by exposing the information to more employees, you are reducing its value as a trade secret.

Finally, were you to have a trade secret inventory containing the actual trade secret information, you would not want to expose it to an external trade secret audit team anyway. The authors are sure that most people who work on such teams are scrupulously honest, but if you were a foreign country or a high-technology firm, whether domestic or international, can you think of a better place to try to insinuate an agent than a trade secret audit team?

The proper course is to build your existing trade secret inventory by collecting trade secret metadata from the employees who need to know the trade secret to do their jobs, but without having them disclose the trade secret information itself, even to internal staff.

Conduct Of The Trade Secret Audit

The trade secret audit team will have their own methods, of course, but there is one thing you should insist on in the conduct of the audit itself. All work should be done on-site, in a set-aside location, with no materials leaving the building. By a set-aside location, we mean a dedicated room or rooms isolated enough to keep outsiders from roaming the halls.

It's a difficult balance. You have sensitive materials and knowledgeable outsiders, and you have to put the two of them together to get a proper trade secret audit done. Either you have to allow sensitive materials to leave the company, or bring outsiders into the company. Taking all the risks into account, it's safer to bring the outsiders in, as long as you control their physical access to the rest of the building. Letting your documentation out of the building, in comparison, is fraught with serious risk.

Assessing The Deliverables

As already mentioned, the deliverables should include a line-by-line report card on the elements of the trade secret program, an evaluation of any deficiencies found, detailed recommendations for correcting the deficiencies, and a forward-looking view of developing threats and possible countermeasures. There is a likelihood that not all of the recommendations will be workable.

A good trade secret audit team will suffer from the same myopia as any other competent specialist. They view their specialty as the be-all and end-all. Some of their recommendations may not be economical to implement, or may not fit the company's culture, or may not allow the flexibility required for the company to carry out its business. One must, for example, tell the people making the product how to use the trade secret information in carrying out their jobs. Or it may be considered more effective to carefully screen the shop-floor janitors than to remove all evidence of the trade secret information every evening before the janitors arrive.

How best to balance the economies, the cultural fit, and the workability of the trade secret audit team's recommendations is left to the company's internal staff, such as the in-house counsel. They know the company better than the outside trade secret audit team, and only they can determine the most common-sense path going forward.

Summation

Periodic trade secret audits are required to ensure the company's program and its execution are sufficient to protect the company's information assets. These audits should be done on-site by an outside trade secret audit team under the direction of counsel. The recommendations of the trade secret audit team must be weighed in light of the company's economics, culture, and work processes in determining how to proceed going forward.

PART IV: THE FUTURE

We should all be concerned about the future because we will have to spend the rest of our lives there.

— Charles F. Kettering

Chapter 16
The Defend Trade Secrets Act of 2016

Executive Summary

The Economic Espionage Act was amended in 2016 to add a civil cause of action, allowing civil suits for trade secret misappropriation to proceed under federal law. The EEA is built upon and echoes the UTSA used in the state courts, but provides federal court procedural advantages in protecting trade secrets in the interstate and international environment most corporations operate in today. The EEA amendments include *ex parte* seizure of stolen trade secret information, allowing swift action to prevent the trade secret information being transmitted or transported outside of the United States and beyond the reach of its courts..

The Economic Espionage Act of 1996

We have previously mentioned in passing the Economic Espionage Act of 1996 (EEA). The EEA had been a criminal statute, making it a federal crime to steal trade secrets. As a criminal statute, its use to protect trade secrets was not available to companies: the EEA could only be used by federal prosecutors. The standard of proof is the "beyond a reasonable doubt" standard of criminal prosecution. Finally, the available remedies — fines and imprisonment — were not the remedies of injunctive relief and monetary damages most important to the corporate trade secret owner. The combination of the stronger criminal burden of proof and the limited resources of federal prosecutors limited EEA prosecutions. As a result, there were less than a dozen EEA prosecutions brought per year, while there have been well over a thousand civil trade secret lawsuits filed each year in the new millennium.

The text of the EEA is provided in Appendix A.3. The definition of a trade secret is largely taken from the UTSA, but has been expanded to make its application to modern computer technology clearer:

> the term 'trade secret' means all forms and types of financial, business, scientific, technical, economic, or engineering information, including patterns, plans, compilations, program devices, formulas, designs, prototypes, methods, techniques, processes, procedures, programs, or codes, whether tangible or intangible, and whether or how stored, compiled, or memorialized physically, electronically, graphically, photographically, or in writing if —
> (A) the owner thereof has taken reasonable measures to keep such information secret; and
> (B) the information derives independent economic value, actual or potential, from not being generally known to, and not being readily ascertainable through proper means by, the public....

Again, we see the four critical elements that we saw in the UTSA: 1) information that provides economic benefit 2) from not being generally known 3) or readily ascertainable and 4) that the owner has taken reasonable measures to keep secret.

Similarly, violations of the EEA mirror the definition of misappropriation in the UTSA:

> (1) steals, or without authorization appropriates, takes, carries away, or conceals, or by fraud, artifice, or deception obtains a trade secret;
> (2) without authorization copies, duplicates, sketches, draws, photographs, downloads, uploads, alters, destroys, photocopies, replicates, transmits, delivers, sends, mails, communicates, or conveys a trade secret;
> (3) receives, buys, or possesses a trade secret, knowing the same to have been stolen or appropriated, obtained, or converted without authorization;

Section 1831 of the EEA addresses theft of trade secrets to "benefit any foreign government, foreign instrumentality, or foreign agent", while Section 1832 addresses theft of trade secrets for the "economic benefit of anyone other than the owner thereof". The reason for the separation is two-fold. First, federal law must address a federal issue. For Section 1831, it is the involvement of a "foreign government, foreign instrumentality, or foreign agent", while Section 1832 applies to theft of a trade secret "that is related to a product or service used in, or intended for use in, interstate or foreign commerce." Second, the EEA provides different penalties for Sections 1831 and 1832. Economic espionage for a foreign entity entails stiffer penalties than trade secret theft for a domestic entity.

What the EEA and the UTSA Were Missing

As a federal criminal statute, the EEA missed the mark for the trade secret owner whose information has been stolen:

1) The case must be proven beyond a reasonable doubt, rather than by a preponderance of the evidence. As trade secret thefts are always clandestine, the evidence to prove the case to

the higher standard may not exist.

2) The case must be brought by a federal prosecutor. That is, the trade secret owner had no control over the criminal prosecution or even whether there will be an indictment.

3) There was no provision in the EEA for civil damages or injunctive relief. As a criminal statute, it included penalties in the form of fines and imprisonment, which certainly deter trade secret theft, but they are of little benefit to the injured party.

4) There was no provision in the EEA for a trade secret owner to move quickly to seize stolen trade secret information before it is transferred to another country beyond the reach of U.S. criminal law.

As a state civil statute, the UTSA also misses the mark for the trade secret owner whose information has been stolen:

1) Attorneys on either side of a trade secret misappropriation lawsuit in state court cannot issue valid subpoenas across state lines without the involvement of courts in both states. This process may involve several court hearings and may take months to resolve. The immediacy of irreparable injury in trade secrets cases makes this a huge problem. A trade secret once lost is lost forever.

2) The UTSA lacks enforcement authority against interstate actors. Collecting a judgment for monetary damages or seeking enforcement of a court order or injunction across state lines is another huge problem.

3) Similarly, the UTSA lacks any enforcement authority against foreign actors. The state courts provide no means of collecting a judgment or enforcing a state court order or injunction against a foreign actor or organization.

4) The UTSA is a state statute without the imprimatur and prestige of a national federal statute. There are federal causes of

action for patents, copyrights and trademarks but not for trade secrets. Even worse, two states—New York and Massachusetts—have not enacted the UTSA. Fortunately, however, New York and Massachusetts are subject to the Restatement (Third) of Unfair Competition that harmonizes U.S. common law with the UTSA, thereby bringing both New York and Massachusetts into the UTSA framework, albeit indirectly.

Some of these issues had been addressed in specific cases by a civil cause of action under the Computer Fraud and Abuse Act (CFAA). The CFAA prohibits a person from using a computer without authorization or in excess of authorization so as to cause damage or loss. There is original jurisdiction in federal court for CFAA violations and the UTSA can be joined to the CFAA cause of action as a "pendent" state law claim. However, the CFAA has a tortured history and there are conflicting decisions that restrict the application of the CFAA to trade secret misappropriation claims. Further, there is a $5000 jurisdictional amount that has been construed to require physical damage to the computer systems in some jurisdictions. In short, the CFAA has not proved to be a satisfactory alternative to a UTSA cause of action for misappropriation of trade secrets in a computer environment.

Similarly, diversity of citizenship — plaintiffs and defendants resident in different states — can be used to establish federal subject matter jurisdiction and file a UTSA case in the federal courts. However, in most cases, there is at least one defendant that resides in the same state as one of the plaintiffs and this destroys diversity of citizenship and deprives the federal court of subject matter jurisdiction.

Given the increasingly interstate and international nature of trade secret cases in the modern economy, and the immediacy of harm in trade secret cases, a solution was required to protect the information assets of the United States economy.

The Defend Trade Secrets Act of 2016

One of the authors proposed two critical amendments to the EEA in the John Marshall Law School Review of Intellectual Property Law in 2008, providing a solution to these issues that caught the attention of lawmakers. This resulted in the Defend Trade Secrets Act of 2016 (DTSA), a bi-partisan and bi-cameral proposal that passed the Congress and was signed into law in early 2016. The full text of the DTSA is contained in Appendix A.4.

The solution lies in two major changes to the original EEA:

1) Adding a federal civil cause of action to the EEA to provide the victims with access to the federal courts in a trade secret misappropriation lawsuit with the full machinery of the federal courts, including national service of process and extraterritorial jurisdiction, and no preemption of the state law UTSA claims.

2) Adding special provisions for *ex parte* seizure orders to preserve the evidence of trade secret misappropriation and to prevent the propagation or dissemination of the trade secrets that are the subject of the action.

The new civil cause of action is Section 1836 of the DTSA, which reads:

> "An owner of a trade secret may bring a civil action under this subsection if the person is aggrieved by a misappropriation of a trade secret that is related to a product or service used in, or intended for use in, interstate or foreign commerce."

The definition of trade secret in the DTSA is unchanged from the UTSA-derived version in the original EEA, while the definitions of misappropriation and improper means are taken directly from the UTSA. This means that the existing body of legal precedent built up in UTSA trade secret cases applies to the application of the DTSA in federal cases.

Benefits of the DTSA

The benefits of the DTSA are hard to overstate:

1) The DTSA provides a uniform trade secret protection regime across the United States.

2) The DTSA provides statutory trade secret protection for the first time in New York and Massachusetts.

3) The DTSA is a civil cause of action, which must be proved by a preponderance of the evidence.

4) The DTSA plaintiff is in control of if and when suit is brought, how it is argued, and how it is settled.

5) The DTSA provides the full range of civil remedies contained within the UTSA.

6) The DTSA provides national service of process, allowing evidence to be discovered across state lines, and extraterritorial jurisdiction to conduct occurring outside the United States.

7) The DTSA enacts special procedures for *ex parte* applications to the Court to preserve evidence and to prevent the propagation or dissemination of trade secrets outside the United States, allowing the trade secret owner to move quickly to seize stolen trade secret information before it is transmitted or transported out of the country.

8) The DTSA provides enforcement of court orders and judgments against interstate actors.

9) The DTSA provides federal means of enforcement of court

orders and judgments involving conduct occurring outside the United States if the offender is a natural person who is a citizen or permanent resident alien of the United States or an entity organized under the law of the United States or a State or political subdivision of the U.S.

10) The DTSA retains the EEA in all other respects, with little or no change to the definitions of trade secret, misappropriation, and improper means from the UTSA, thereby continuing to benefit from the 36-year history and case precedent under the UTSA at all levels of the U.S. judiciary (both state and federal courts) which, in turn, forms the bedrock of the EEA.

11) The DTSA will make it unnecessary to invoke CFAA claims or diversity of citizenship in order to bring trade secrets cases in federal courts.

RICO Provision

The DTSA also amends the EEA to add Section 1831 and Section 1832 as predicate acts under the civil Racketeering and Corrupt Organization Act (RICO) 18 U.S.C. Section 1961 and 1964. This will now provide an avenue for bringing a civil cause of action for RICO violations based upon two or more predicate acts of trade secret theft within a 10–year period. The civil RICO statute now, in conjunction with the EEA, provides an avenue for litigating trade secret "conspiracy" cases involving foreign actors and others who are co-conspirators in the theft of trade secrets even if another co-conspirator committed the trade secret violations. This is a huge deterrent to foreign trade secret theft because the location of the bad actors cannot forestall the prosecution of the RICO cause of action in U.S. courts. See Chapter 17 for an in-depth discussion of RICO in the trade secret context.

The DTSA in Practice

How does the DTSA change trade secrets law? What changes does it make to the litigation of trade secrets cases? And what changes does it make to the first fifteen chapters of this book?

The important changes to trade secrets law in the DTSA are procedural not substantive changes. The DTSA short-circuits a lot of procedural roadblocks, such as court hearings in multiple states on interstate subpoenas, venue arguments about diversity of citizenship, and motions to dismiss CFAA claims. These procedural issues have made trade secret litigation more complex, more lengthy, and more costly, and merely served to delay or deny the plaintiff's right to seek emergency or preliminary injunctive relief and to a full trial on the merits.

The DTSA allows the law of trade secrets to continue to flourish in the 21st Century and assists victims of trade secret misappropriation to prevent the loss of trade secrets overseas. The DTSA does not change any substantive aspects of trade secret law, which dates back in the United States almost 200 years. With the bedrock definitions of trade secret, misappropriation, and improper means remaining unchanged in the DTSA and taken with little or no change from the UTSA, the federal courts are able to apply decades of decisions under the UTSA to DTSA civil actions. All of the discussions in previous chapters still apply. Plaintiffs still have to prove the elements of existence, ownership, notice, and access in addition to use and damages. Existence of a trade secret is determined from a very similar definition, and a consideration of the six factors from the Restatement (First) of Torts is still the criteria for determining whether an alleged information asset qualifies as a trade secret asset under the DTSA. The right to injunctive relief and damages remains unchanged, as do other provisions of the UTSA that are incorporated into the EEA and retained in the DTSA.

Finally, the trade secret owner retains the option of filing the trade secret misappropriation claim in state court under the UTSA rather than federal court under the DTSA. This may be the preferred venue for smaller cases, or smaller litigants, when interstate or international issues are not involved, or for a variety of other reasons, avoiding the increased expense of litigating under the federal rules.

Summation

The Defend Trade Secrets Act of 2016 added a civil cause of action to the Economic Espionage Act. In effect, the DTSA elevated the Uniform Trade Secrets Act to federal status, using the same or similar definitions and remedies as the UTSA. The DTSA thus preserved existing case law, while simplifying procedural issues for trade secrets cases with interstate or international issues. The UTSA statutes in the individual states remain available to plaintiffs for whom interstate and international issues are not in play.

Chapter 17
The RICO Provisions of the DTSA

Executive Summary

The Defend Trade Secrets Act of 2016 made theft of trade secrets a predicate offense under the RICO statute. RICO provides a more robust statutory framework to pursue the sometimes complex conspiracies that commit the theft of trade secrets. Treble damages, widened liability, extended limitations, and widened venue and jurisdiction, including extraterritorial jurisdiction, are some of the additional tools available to the trade secret plaintiff under RICO.

The RICO Provisions of the Defend Trade Secrets Act of 2016

There is a lot of commentary about the DTSA's creation of a federal cause of action for trade secret misappropriation and the enactment of *ex parte* seizure provisions. These are two critical amendments that can be traced back to the primary author's law review article in 2008 entitled *Protection of U.S. Trade Secret Assets: Critical Amendments to the Economic Espionage Act of 1996.*

However, there is a much bigger development in the DTSA legislation. Although overlooked by most legal pundits, the DTSA also adds Sections 1831 and 1832 of the Economic Espionage Act of 1996 (the EEA) as predicate criminal acts in Section 1961 of the Racketeering Influenced and Corrupt

Organizations Act (RICO).

Adding Sections 1831 and 1832 to the list of RICO predicate acts activates EEA Section 1837, which extends extraterritorial jurisdiction to conduct occurring outside the United States if (1) the offender is a citizen or permanent resident alien of the United States, or (2) an organization that is organized under the laws of the United States or a State or political subdivision thereof, or (3) an act in furtherance of the offense was committed in the United States.

This means that the target list of potential defendants in a civil action now includes RICO defendants, which extends civil liability for trade secret theft well beyond the statutory limitations of a UTSA claim.

Racketeer Influenced and Corrupt Organizations Act, Title IX of the Organized Crime Control Act of 1970 (RICO)

The Organized Crime Control Act (OCCA) was passed in 1970. The OCCA attacked organized crime in the United States by strengthening the legal tools available to both prosecutors and victims of organized crime. The RICO Act created four new criminal offenses and a new civil cause of action for "[a]ny person injured in his business or property by reason of a violation" of those offenses.

As Supreme Court Associate Justice Blackmun observed in 1983: "The legislative history clearly demonstrates that the RICO statute was intended to provide new weapons of unprecedented scope for an assault upon organized crime and its economic roots."

The rampant rise in foreign economic espionage and trade secret theft parallels the rampant rise in organized crime years ago. Today, Former NSA Director General Keith Alexander says cyber-espionage is causing the "the greatest transfer of wealth in history." Symantec places the cost of intellectual property theft in the U.S. economy at over $250 billion a year. Adding EEA Sections 1831 and 1832 as RICO predicate acts now opens entire new avenues for victims of economic espionage and trade secret theft.

RICO punishes enterprise criminality. The criminal offenses created by RICO target a pattern of racketeering activity committed by, through, or against an enterprise. Organized crime is a special category of crimes involving the activities of organized criminal groups in relation to an enterprise. RICO is the key enforcement tool used by the FBI to expand criminal liability across multiple members of a criminal enterprise and to capture RICO defendants and conspirators that often reside in safe haven countries.

EEA Sections 1831 and 1832 fall naturally into a RICO organized crime framework. Without RICO offenses, foreign governments can operate by or through foreign instrumentalities, foreign agents, or U.S. persons without exposure to civil liability under U.S. law.

The RICO statute defines racketeering activity to encompass over 50 federal and state offenses, now including EEA Sections 1831 and 1832, as RICO predicate acts. Each of these offenses is itself a crime, but RICO defines these offenses as racketeering activities and predicate criminal acts to establish enterprise liability for RICO defendants. A defendant can face criminal or civil liability under RICO even if the defendant has not been convicted of the underlying RICO predicate acts.

The elements of a RICO civil cause of action appear intricate and complex at first blush, but are really not complicated: The plaintiff sues the RICO defendant for conducting the affairs of a distinct and separate RICO enterprise through a pattern of racketeering activity. A pattern of racketeering activity requires at least two Section 1961 predicate criminal acts committed within 10 years of each other. The United States Supreme Court has also superimposed the judicially-created requirement that the pattern of racketeering activity include the existence or threat of continued criminal activity.

The RICO enterprise is defined as "any individual, partnership, corporation, association or other legal entity, and any union or group of individuals associated in fact although not a legal entity." The RICO enterprise is thus separate and distinct from the RICO defendant. RICO does not target the RICO enterprise. RICO targets the RICO defendants who corrupt legitimate enterprises through a pattern of racketeering activity.

The origin of the RICO statute is organized crime. The statute reaches RICO defendants who may not have blood on their own hands, but who operate or manage or corrupt the RICO enterprise by engaging in a pattern of racketeering activity. RICO exposes the "mastermind" or "Mr. Big" or the "Godfather" to criminal or civil liability for engaging in a pattern of racketeering activity while sitting in a three-piece suit in an executive office suite, or perhaps serving as a high-ranking official working for a foreign government.

The critical point for litigators is that the defendant may violate RICO even if he or she does not commit the underlying crimes, does not direct someone else to commit the underlying crimes, and does not conspire with someone else to commit the underlying crimes. A defendant can violate RICO simply by operating or managing an enterprise through a pattern of racketeering activity.

RICO Jurisdiction and Venue Provisions

RICO has jurisdiction and venue provisions that favor plaintiffs. Section 1965(a) provides that venue is proper in any district in which the defendant resides, is found, has an agent, or transacts his affairs. Section 1965(b) provides that in any RICO action "in which it is shown that the ends of justice require that other parties residing in any other district be brought before the court, the court may cause such parties to be summoned, and process for that purpose may be served in any judicial district of the United States by the marshal thereof." Finally, Section 1965(d) provides that all other process may be served on any person in any judicial district in which such person resides, is found, has an agent, or transacts his affairs.

Combining these three statutory provisions, the courts have approved national service of process under Section 1965(b) and Sections 1965(d), venue under Section 1965(a), and the haling into a federal district court of out-of-district RICO defendants under Section 1965(b) if venue is proper as to at least one RICO defendant under Section 1965(a).

In trade secrets cases, jurisdiction and venue can be big issues for the plaintiff. Defendants may be spread across multiple jurisdictions, or the offense may have been committed piecemeal in multiple jurisdictions. Under the UTSA and even the DTSA, establishing venue and jurisdiction may be a big court battle before any substantive issues are raised.

But in a RICO action, venue and jurisdiction are proper in any U.S. court where any of the defendants lives or conducts business, and all the other defendants can be summoned into that court to answer the plaintiff's claims.

RICO Extraterritorial Jurisdiction

The RICO enterprise can encompass foreign governments, foreign instrumentalities, and foreign agents. There is also applicability of Sections 1831 and 1832 to conduct outside the United States. Therefore, RICO applies to extraterritorial conduct because the predicate Section 1831 and 1832 offenses themselves apply extraterritorially. This was the holding in the recent U.S. Supreme Court decision in *RJR Nabisco, Inc. v. European Community* decided on June 20, 2016.

RICO Widens the Scope of Liability for Trade Secret Theft

The UTSA and DTSA statutory causes of action impose liability on the direct misappropriator who wrongfully acquires, discloses or uses trade secrets and a very limited class of third parties who know or have reason to know that the trade secret was acquired by improper means. However, in a global economy, the threat and continuing misappropriation of trade secrets involves various actors involved in sophisticated multinational schemes, conspiracies and criminal enterprises to steal trade secrets. Often the bad actors involve foreign governments, foreign agents, and foreign instrumentalities far removed from the actual misappropriators, but who actively operate or manage the criminal enterprise to steal or attempt to steal trade secrets. Utilizing Section 1831 and 1832 predicate offenses, all the participants and co-conspirators can now be sued in a RICO civil action in a U.S. District Court.

The EEA Section 1831 and 1832 offenses — now RICO predicate acts — capture a much wider universe of bad acts using the stolen trade secrets: "copies, duplicates, sketches, draws photographs, downloads, uploads, alters, destroys, photocopies, replicates, transmits, delivers, sends, mails communicates, or conveys a trade secret." Sections 1831 and 1832 are perfect predicates for a pattern of racketeering activity. There are over 25 wrongful acts to choose from (in either Section 1831 or Section 1832) and 75 wrongful acts (in either Section 1831 or Section 1832) when "attempts" and "conspiracies" to commit offenses are included.

The whole point of RICO was to eliminate the effectiveness of cut-outs and plausible deniability on the part of the remote decision-makers who run corrupt organizations, and these remote decision-makers can now be held liable for the corrupt organization's participation in the theft of trade secrets.

RICO Statute of Limitations and 10-Year Discovery Period

The RICO statute of limitations is four years. The UTSA and DTSA statute of limitations is three years. A continuing misappropriation constitutes a single claim under the UTSA and the DTSA. However, RICO does not consolidate predicate acts into one act, so a continuing theft of trade secrets will continue the pattern of racketeering activity and extend the statute of limitations.

In addition, a RICO civil cause of action supports a 10-year discovery period because a pattern of racketeering activity is defined by the statute as two predicate acts committed within 10 years of each other. Placing litigation holds on RICO defendants dating back 10 years, which they must obey or face spoliation claims and judicial sanctions, is also a powerful weapon in RICO litigation. The district courts have uniformly held that a time-barred predicate offense may still serve as a RICO predicate offense for an otherwise timely RICO claim.

RICO Treble Damages, Costs and Attorney's Fees

RICO civil remedies are clear-cut and mandatory. If a plaintiff proves $100,000 in damages resulting from a RICO violation, then the plaintiff gets $300,000 in damages plus its costs and attorney's fees.

In contrast, the plaintiff in a UTSA action must prove that the defendant's trade secret misappropriation was willful and malicious to recover increased damages and attorney's fees, and such additional awards are discretionary with the trial court. In a DTSA action, the plaintiff cannot obtain increased damages and attorney's fees if the trade secret owner has not complied with the whistleblower notification provisions in the DTSA. And, like the UTSA, an award of increased damages and attorney's fees is discretionary with the trial court in a DTSA action.

There is no such discretion left to the trial court in a successful RICO action. Treble damages plus costs and attorney's fees are mandatory under RICO.

Summation

The listing of violations of Sections 1831 and 1832 of the Economic Espionage Act under Section 1961 of the Racketeering and Corrupt Organizations Act provides the means for the trade secret plaintiff to go after the entire corrupt organization that stole his trade secrets and profited from the theft. Designed to go after mob kingpins and organized crime syndicates, RICO is particularly well suited to pursue the domestic and international organizations that target, steal, and profit from advanced trade secret technologies.

Chapter 18
COVID And Trade Secrets

Executive Summary

The COVID pandemic from early 2020 into 2022 changed many things. Many of these changes are proving long-lasting, a change in approach that has shown endurance even as the pandemic itself receded in severity. These changes made trade secret security even harder, and put trade secrets at greater risk than ever before. We expect these changes to persist.

The Response to the COVID Pandemic Changed Everything

The response to the COVID pandemic significantly worsened the landscape for protecting trade secrets. Some of this was due to direct corporate actions, and some of it was part of the greater impact on the economy and the business environment. This worsened landscape has effects that have persisted and will shape the future of the trade secret environment.

Among these effects are the increased work-from-home, reduced social attachment to the workplace, reduced employee loyalty, vaccine and mask mandates, increased employee turnover in the workplace, and decreased corporate insistence on company-owned devices. We will look at each of these areas in turn.

Increased Work From Home

One response to the COVID pandemic was corporations allowing significantly increased work from home. Employees who did not come into the office were at significantly reduced risk of contracting the virus or infecting others with it. As a response to a viral infection making the rounds, it was certainly an understandable strategy.

The strategy was widespread in U.S. companies. Of U.S. workers who say their jobs can mainly be done from home, the percentage actually working from home all or most of the time went from 23% before the COVID outbreak to 71% by October 2020. That number backed off to 59% by February of 2022, but remained much higher than pre-COVID levels.

The employees most likely to be permitted greater freedom to work from home were primarily information workers. Someone who runs a machine on an assembly line cannot work from home. Someone who spends their work time generating, manipulating, and transmitting information, including proprietary information like trade secrets, was the most likely candidate for increased work at home.

At the same time, increased work from home reduces the awareness of the need for information security. We are all in our off-time, free agents, and most of that off-time is spent at home. Employees are also aware that in the office they are agents of the company, working toward the company's goals.

Working from home risks confusing the two, free agency and company agency. The employee is less reminded by his location of his obligations to the firm, and may be more likely to elevate his own priorities and goals, such as personal aggrandizement at the cost of the employer's security in its proprietary information.

At the same time, at home the employee is surrounded by people who are not included in the employer's circle of confidentiality, such as children — even adult children — spouses, and roommates who may also be working or attending school at home. The risks of others overhearing company proprietary information are greater, and that may be without any appreciation of the confidentiality or security that applies to the information.

Those children, spouses, and roommates are not part of the company's workforce, and are not privy to the company's security and information-handling briefings and training. The risk of disclosure, malicious or not, is therefore heightened.

Reduced Social Attachment to the Workplace

Another factor in employee's protection and proper handling of proprietary information is their social attachment to the workplace. COVID significantly reduced the opportunities and benefits of employees' social contacts at work.

Social distancing, work from home, virtual meetings, and similar factors reduced the healthy social interaction of employees with each other. Employment is more than hiring oneself out to carry forward the employer's goals. For most people, it is the primary social activity they engage in. They have friends at work. They work with them during the day, they often socialize with them at lunch, during the evening, or on weekends.

One of the restraining influences preventing the theft or misappropriation of proprietary information is the sense that they would be letting their friends at work down. In addition to the element of personal betrayal is the sense that they would be stealing from them, and reducing the viability of their employer. They are putting their friends' continued employment at risk, by endangering the profitability of the company.

A strong social attachment to the workplace, in the form of their friendships with their fellow employees and management, reduces the risk of trade secret misappropriation. Alienation from fellow employees, especially management, is one of the indicators in our Domestic Trade Secrets Misappropriator Profile in Table 10-1. But the steps taken in response to the COVID pandemic were all about keeping employees apart.

Social distancing, electronic meetings from one's desk, work from home, limiting contact outside of work by closing gathering places all were inimical to social attachment to one's fellow employees. This served to put trade secrets at further risk.

Reduced Employee Loyalty

One of the impacts of the COVID pandemic was a disrupted marketplace for many products and services. Many companies were forced to lay off workers due to reduced revenues. These layoffs served to reduce employee loyalty to the company.

Layoffs — particularly multiple rounds of layoffs over months or years — shake employee confidence in the stability of their employment. Many employees wonder, "Am I next?" Often the employee response to layoffs is to begin a search for new employment, initially as a defensive move to shorten the time between jobs.

As this search proceeds, however, the employee is shown new opportunities, encounters new environments, and begins to consider alternatives. They ponder the wisdom of leaving before the next round of layoffs, of getting out while the getting is good. It is natural that their loyalty to the existing employer begins to wane as the employee mentally prepares for the potential of a new job with a new employer.

There may also be an element of bitterness at this point. The employer is certainly taking care of their bottom line, laying off employees — even long-term employees — to ensure the survival of the company. Employees may begin to feel that they, too, should put their emphasis on their personal well-being over the rules and rights of the company.

In this environment, trade secrets are more at risk.

Vaccine and Mask Mandates

Many employers, once sufficient supplies of masks became available, mandated that their employees wear masks against COVID. Some of these mandates came from federal or local governments, some were at the employer's own discretion.

Many employers, once vaccines began rolling out in February 2021, mandated that their employees be vaccinated against COVID. Again, some of these mandates came from federal or local governments, some were at the employer's own discretion.

Some employees had serious issues with vaccine and mask mandates, ranging from the medical -- due to their existing medical history -- to the religious. Other employees had serious issues with employers who did not mandate either the vaccine or masks, due to being forced to work with the unvaccinated or the unmasked. Either way a company decided, it lost, in that some of its employees were guaranteed to be unhappy with its position on the matter.

Employees left their employers for both of the above

reasons, though more were forced out by vaccine mandates, especially in the healthcare and transportation industries. None of these actions on the part of companies struggling to remain profitable in the economic turbulence of the pandemic were without impact on employee morale or loyalty to the company. Many of the employees forced out by failure to comply or their own fears about the unvaccinated or unmasked harbored bitterness to their former employer on their separation from the company.

Increased Employee Turnover in the Workplace

It is not surprising, then, that all of these elements led to increased employee turnover in the workplace. Employees left their employers at record rates in 2021, to the point that it was called the Great Resignation. Higher turnover rates than those prevalent prior to the COVID pandemic have since persisted.

A further incentive to change jobs was the economic disruption resulting from the pandemic. Such economic disruption results in employment openings as some companies who were positively affected -- such as those providing medical and safety equipment like plastic screens -- staffed up, and other companies forced out those employees who would not comply with the company's policy with regard to safety measures.

Further employment openings resulted from parents leaving the workforce to care for their children at home due to many schools being shuttered.

An employee with knowledge of the company's trade secrets has a limited prospect for the improper use of the information as long as they stay with the company. With increased employee turnover, however, more former employees had the opportunity and the incentive to misuse their prior employer's proprietary information in their new jobs.

Decreased Corporate Insistence on Company-Owned Devices

The increase in work from home tempted companies to reduce their insistence on the use of company-owned devices. Employees were in many cases permitted to use their home computers to log in to the company to perform their work-from-home tasks. This Bring-Your-Own-Device (BYOD) approach is fraught with danger.

In Chapter 7, we recommended against BYOD: "Permitting such use of personal devices is fraught with risk: the company should provide and pay for all electronic devices necessary for the employee's work tasks, and strictly prohibit the use of personal electronic devices for any company business."

All devices used by the employee in performing work tasks, particularly the generation, handling, and transmission of proprietary information, should be on Company Owned, Personally Enabled (COPE) devices. The COVID pandemic did not change this recommendation, and it did not mitigate the risks we warned against in Chapter 7.

The incentive for companies to switch from a COPE strategy to a BYOD strategy in the face of the pandemic is understandable. Shifting the cost burden of support devices to the employee certainly saves money for the company, particularly at a time of economic disruption and uncertainty.

There was also the convenience factor. Companies whose employees primarily use desktop computers in the office would have had to purchase large numbers of laptops for work-at-home employees. In addition to the expense of the purchase, the configuring of all these company-owned devices by the corporate IT staff may have been a large logistical problem.

Nevertheless, the dangers of BYOD remain unchanged from our discussion in Chapter 7. The risks to trade secrets remains. Moving away from COPE, even in the face of the COVID pandemic, was a mistake, and we recommend correcting this mistake wherever it was made.

The Increased Risk to Trade Secrets Is Real and Persistent

Any one of the changes above caused by the response to the COVID pandemic would significantly increase the risk of trade secret misappropriation. Taken together, they represent a veritable tsunami of risk, and, as with a tsunami, it will take a long time for the water to recede.

We expect some of the changes brought about by the response to the COVID pandemic will persist. People are resistant to change habits and practices without a big push. The COVID pandemic provided the big push to change many habits and practices, but, now changed, many of those changes will endure.

We have seen some attempts by companies to roll back the changes, at least to some extent. Vaccine and mask mandates are now gone in most locations, healthcare facilities being a notable exception. Work-from-home policies are now being reversed in some companies. There is a lot of resistance to that move, as work-from-home has become popular among employees. The disruption to employee morale and loyalty of layoffs has subsided as employers have stabilized their workforce levels.

Other changes are occurring on their own. Employee

turnover is falling slowly. Decreased work-from-home is increasing the social attachment to the workplace. Smart companies are reconsidering their acquiescence to the use of employee-owned devices for work-related tasks.

All of these reversions toward prior levels, though, will not go back to pre-COVID levels. They will stabilize at some higher level into the future. The risk to trade secrets remains elevated. It is imperative that the trade secret owner be aware of these increased risk levels, and take appropriate measures to mitigate those increased risks.

Summation

The responses to the COVID pandemic unwittingly raised the risks to trade secrets at most companies. Work from home, reduced social attachment to the workplace, increased turnover, and the increased use of employee-owned devices were particularly powerful in increasing these risks. We expect these risks to be persistent, and caution companies to evaluate and mitigate these increased risks in their individual circumstances. In particular, automated trade secret asset management, which we will cover next, can help address these increased risks.

Chapter 19
Automated Trade Secret Asset Management

Executive Summary

The automated management of trade secret assets has long been considered impossible. It is true that trade secrets cannot be managed in exactly the same manner as other company assets, but automated management is possible if one does not lose sight of the essential nature of trade secrets and the measures necessary to protect and defend them.

Approach To Automation

Having considered in the first seventeen chapters of this book the nature of trade secrets, the means by which they can be either lost or defended, and the more advanced concepts of categorization, identification, classification, valuation, and assetization, it is left to consider how much of best-in-class trade secret management practices can be automated.

Like other corporate processes, such as payroll, inventory management, and customer relationship management, automating trade secret asset management practices results in cost savings, timeliness of information for decision makers, and overall efficiency of the organization. But in the case of automated trade secret asset management, there is an even greater benefit to the company than any of these.

Automation of trade secret asset management enables the company to be proactive. Employees can be better informed of their obligations of confidentiality and non-use, perhaps forestalling misappropriation and litigation altogether. Litigation to defend trade secrets can be quicker, cleaner, and more strategic. Security measures can be tailored, and respond to changes in the trade secret landscape of the company in real time. Automation of the trade secret management process can do away with those emergency fire drills with which information managers have become increasingly familiar.

For these and many other reasons, the authors have been developing and using methods for automating the trade secret asset management process since before the turn of the millennium

The Five Essential Elements of Automation

There are five essential elements of an effective trade secret asset management software system:

1 - Taxonomy. The trade secrets must be classified into some easily understood taxonomy. Trade secrets existing as a large blob of undifferentiated knowledge is unacceptable. The classification system must be easy to understand so employees need not be trained to use it.

2 - Scoring. The trade secrets must be scored by some method that reflects legal standards and is easy to use. All trade secrets are not created the same. Some are better than others. The legal standard for trade secrets involves a six-factor litmus test from the Restatement (First) of Torts that can be used to determine whether an alleged information asset qualifies as a statutory trade secret

3 - Metadata. The automated trade secret asset management system should only store metadata — data about the trade secrets — not the trade secrets themselves. Were the system to store the trade secret information, it would become a security risk. Further, putting the trade secret information into the system exposes more people to the trade secret information and reduces the score for one of the six factors, to what extent the trade secret is known inside the company. A method of asset management should not reduce the value or security of the assets themselves. Like a library card catalog, which contains no books, the trade secret asset management system should contain no trade secrets.

4 - History. The automated trade secret asset management system must retain all the metadata about trade secrets, and be able to reproduce the metadata in effect at any time in the past. The period of interest in a trade secret litigation may be two or three years in the past. One cannot litigate a case based on today's metadata, one must litigate based on the metadata in effect during the period of interest. Further, the historical treatment of the trade secret — proving reasonable security measures have been taken — will be an issue at trial. The system must retain all the old versions of any trade secret metadata, and be able to reproduce the metadata as it existed at any prior time.

5 - Proof. The automated trade secret asset management system must contain methods for proving that historical metadata is accurate and contemporaneous to the period of interest, and is not a construct prepared for trial.

Taxonomy

We discussed the SFP method for categorization of trade secrets in Chapter 12. This method is particularly well suited to automation. Once the possible Subjects, Formats, and Products within the company have been defined, specifying the three-dimensional space of trade secrets, assigning each trade secret to an SFP is simply a matter of selecting the appropriate Subject, Format, and Product from three drop-down boxes. This is no more complicated than selecting the expiration month and year of your credit card for a payment to an on-line retailer.

Once sorted into SFPs, the trade secrets can be managed in bundles, whether by individual SFP, or by selecting all of the trade secrets on a distinct Subject, in a given Format, or relating to a specific Product or family of Products.

Scoring — Existence

We noted in Chapter 3 that trade secrets can only be validated in litigation, and that the EONA proofs were critical to prove trade secret rights. The Existence, Ownership, Notice, and Access proofs must succeed in court for the trade secret owner to prevail in a complaint of trade secret misappropriation. So what can we do with computer automation to make these proofs easier?

The Existence proof depends on the six factors of a trade secret, described in detail in Chapter 3. If we score each of the factors on a one-to-five scale with regard to each trade secret, as described in Chapter 12, we can perform a calculation combining the six factors in order to come up with a Defendability Factor for each trade secret. The Defendability

Factor represents the likelihood that the alleged trade secret will be judged an actual trade secret in court. This allows counsel to make educated choices of which alleged trade secrets to assert in a future trade secret misappropriation case.

While there may be a thousand trade secrets stolen in a given misappropriation, alleging a thousand trade secrets in court is poor litigation strategy. As the jury must make two determinations for each alleged trade secret — is it in fact a trade secret, and was it misappropriated by defendant — a thousand alleged trade secrets asserted at trial would require two thousand jury decisions. In practice, the court will not allow this to proceed. Counsel for both parties will likely be instructed to "pick your best ten to argue." Asserting a thousand trade secrets has thus allowed defendant's counsel to pick your worst ten trade secrets to defend against. Better, given that plaintiff is "first to the battle," is to pick the "high ground": assert your ten best trade secrets — those ten at issue with the highest Defendability Factor — and give defendant's counsel no easy avenue.

Intersection Of Taxonomy and Scoring

The previous section specified the one-to-five ranking of each trade secret on each of the six factors. That may be an intimidating prospect if millions of pieces of information are involved. But the intersection of SFP classification — the taxonomy — and six factor identification — the scoring — provides relief.

In practice, we have found that all of the trade secrets within a given SFP have similar six-factor rankings.

The importance of this finding cannot be overstated. While

the individual ranking of important and mission-critical information is still recommended, the bulk of the trade secret information within a company can be ranked by ranking the SFPs, not the individual trade secrets.

The importance of taxonomy now becomes clear. Similar trade secrets have been grouped together. It turns out that they can be scored together as well. What began as an intractable problem in information management has become possible, even easy.

Metadata — Leaving The Trade Secrets Where They Are

An automated trade secret management system should be implemented without moving the trade secret information, without exposing it to additional employees, and especially without exposing it to outsiders. There is no need to have interviews with employees, bring in outside consultants, or collect trade secret information in a central repository that constitutes a security target in its own right.

The entry of the SFP and six-factor information about the company's trade secrets should be done by the employees who already have knowledge of the trade secret information, by entering it on a web-page form on the company's internal network. This is why the simplicity and ease-of-use of the SFP classification and one-to-five six-factor scoring of trade secrets is so important. If the employees who already know the trade secret information enter the data, we have not increased the risk to that information by increasing exposure or copying it to additional locations.

Metadata — Security Management

There are two major foci of security management: 1) Is there a security weakness in some places, risking loss (alpha error)? 2) Is security too tight in some places, increasing costs and decreasing the organization's effectiveness and flexibility (beta error)? Automated trade secret management addresses both of these issues.

An automated trade secret management system that contains the six-factor rankings and the Defendability Factor calculations for all the trade secrets in the company can compare the security factor to the Defendability Factor for every trade secret, showing mismatches in the security applied to a trade secret vis-a-vis its defendability in court. Similarly, the system can compare the security factor to the economic benefit factor for all the trade secrets in the company, showing mismatches in the security applied to the trade secret vis-a-vis its economic value to the company.

Secondary analyses of these mismatches by Subject, Format, Product, and location may highlight a systemic problem with security, whether in the handling of certain types of documents, or within a department, a product group, or a facility.

The ability to analyze the security measures applied to all of the trade secret information of the company in minutes, to save costs by reducing beta errors, and to reduce risk of loss by reducing alpha errors, is an additional powerful outcome of automating the trade secret management process.

Metadata and History — Ownership, Notice, and Access

Similarly for the other three EONA proofs, automated

record-keeping and the retention of those records simplify the proof and makes it unassailable.

Ownership of the trade secret information is easy to prove when the originator of the trade secret information is a matter of record and the trade secret assignment signed by that employee is stored in the employee's records. The tracking and security management of the trade secret information over time further proves the company's ownership stake in the information.

The Notice requirement is met when the departing employee is presented at an exit interview with the computer-generated list of trade secrets to which he was exposed during his employment, and this list and the exit interview form signed by the departing employee and a human resources representative is maintained in the employee's personnel files and stored within the system.

Finally, Access to any trade secret is proven by comparing the trade secret's physical, project, and organizational locations within the company over time to the physical locations, project assignments, and reporting structures within which the employee worked during his employment. If the computer database contains the employee's records — which group he worked with, and what projects he worked on, during which periods of his employment with the company — and the trade secret records — which trade secrets were known in which groups, and used in which projects, and the dates of that knowledge and use — then the trade secrets to which the employee has been exposed can be found by automated means. Sorting them by Defendability Factor is a tremendous aid to counsel in determining which alleged trade secrets are most at risk in the incident, and which alleged trade secrets are the best choices to assert at trial.

Putting It All Together — Answering the Motion To Compel

The most important and difficult part of any trade secret litigation is defining the *res*, the "it" analysis. The complaint alleges the theft of a trade secret, so what is "it"? What is the trade secret? This is the foundational question of the complaint, the bedrock on which all further action in this litigation will proceed. It is the most important issue to get right from the start, and many trade secret litigations fail months or years and millions of dollars later for a weakness or failure of this analysis.

Litigation always begins with a complaint and an answer to the complaint, but the very next step in a trade secret misappropriation matter is likely to be Defendant's expedited motion demanding that plaintiff identify with specificity and particularity the trade secrets alleged to have been stolen. Note that this is before discovery, so no fishing expedition can answer this question. Absent automation and a proactive inventory of trade secrets, responding to this motion takes considerable attorney time, and is done under the press of the litigation schedule. This is not the way to build the foundation for a successful litigation to defend one's trade secret rights.

With computer automation, it is another matter. Entering the defendant employee's name in the system, one can immediately obtain a list of the trade secrets to which the employee has been exposed, sorted by Defendability Factor, from which to choose those alleged trade secrets to assert at trial. One can only imagine the dismay of defendant's counsel when the well prepared plaintiff's response to the most difficult and important motion in any trade secrets litigation arrives the very next day.

Proof — The Role Of Hash Codes

Plaintiff has submitted the list of alleged trade secrets it will assert in court, along with the trade secret database information documenting each trade secret: the name, description, six factors analysis, locations, dates, security. How can plaintiff prove that this information is historical, and not prepared in the recent past, dates notwithstanding, in order to pursue and punish an innocent departing employee? Automated trade secret asset management facilitates this process as well.

First, a short note on computer hash codes. A computer hash code is a string of characters that can be generated by a computer from any digital input. It is a mathematical process that is repeatable and one-way-only: the same data will always generate the same hash code, but the hash code cannot be used to regenerate the data. A single computer hash code looks like the following:

a7547117bc74e597fd15068fc76f777059b2f162005f02c550c26847dc3409d5

Each digit is one of sixteen values, from 0-9 and A-F, each of which represents four bits (0000, 0001, 0010, 0011, 0100...). This 64-character hash code thus has 256 bits, and is called a 256-bit hash. A 256-bit hash has 2^{256} values, or 1.16×10^{77} possibilities. That is enough numbers that you could give a unique serial number to every neutron, proton, and electron in the Milky Way galaxy, *a million times over*, and still only use half of them

These properties of computer hash codes have been accepted by the courts, and no court challenge to these basic principles will be allowed.

Now back to the trade secret management system.

Whenever a trade secret is entered into the system, or whenever the data about a trade secret changes, an automated trade secret management system can generate a hash code from that data. That hash code can then be submitted to a time-stamping authority, which records the value and the date. For that matter, the hash code could be printed in the New York Times, as the trade secret data it was generated from is not recoverable from the hash code. The same trade secret data at any later date will result in the same hash code, proving that that data existed as of the date it was recorded, whether that was yesterday or ten years ago. If a single byte, a single character, of the trade secret data is changed, the hash code will not match the historical, recorded hash code.

With 1.16×10^{77} possibilities, the odds of any given hash code matching another are one in a hundred thousand trillion trillion trillion trillion trillion trillion. The time-stamping authority could receive a billion hash codes a year for a million years, and the odds that *any* two of them matched would still be infinitesimal.

Proof — The Role Of Blockchaining

A block chain is a set of entries, each of which is mathematically linked to the one before by hash codes. This is how it works. The nth entry in a blockchain is:

data n, timestamp n, hashcode(haschode n-1, data n, timestamp n)

The next entry in a blockchain contains the new data, the current timestamp, and a hash code created from the last entry's hash code together with the new data and timestamp. It won't take much thought to conclude that the chain can't be tampered with, or the hash codes won't calculate properly.

That's the beauty of blockchaining. We get all the benefits of using hash codes, but we don't need a third-party time-stamping authority. The timestamp is built into the chain. There is no possibility of tampering by anyone, including the trusted third party.

In practice, what we do is take the value of the hash code from the trade secret that's been changed, and put it in the blockchain as the new data. This gives us a unique ID for our information that won't match anything else, ever, but without disclosing the information. We link that into the blockchain, which itself depends on hash codes to operate, and now it and its effective date are bound there like flies in amber. Neither can ever be changed, or the whole blockchain will fail a simple hash check. That means the unique hash code for our data is proved, as of a date certain, and since that hash code is unique to our data, it proves our data was in existence on that date as well.

What Of Reasonable Measures?

As we noted in Chapter 1, paragraph one, a trade secret requires that the information owner take reasonable measures to keep the information secret. Reasonable measures, though, is in practice an evolving target. As trade secrets have gotten more complex, as the storage media used to record trade secret information have gotten more complex, as threats to trade secrets have gotten more complex, and as security measures themselves have gotten more complex, the security measures that are considered reasonable have become more involved. Where door locks might once have been considered enough, burglar alarms, then security cameras, then on-line security monitoring became commonplace. Where computer passwords might once have been considered enough, elaborate network firewalls and intrusion detection are now commonplace.

So with trade secret asset management. Where a manual paper inventory of a company's trade secrets may once have been considered best in class, those days are gone. As automation of trade secret asset management becomes mainstream, we are rapidly coming to a time when failure to have automated trade secret asset management will be considered a failure to take reasonable measures.

What About Europe?

On 8 June 2016 following a proposal from the European Commission, the European Parliament and the Council adopted a Directive that aims to standardize the national laws in EU countries against the unlawful acquisition, disclosure and use of trade secrets. As this book goes to press, the deadline on individual nations for compliance with the European Directive is closing in. European companies will need to be in compliance with these laws to have the benefits of trade secret protection under law.

One requirement for an automated trade secret asset management system, therefore, is that it meet the requirements of the European Union directive in identifying, classifying, and documenting trade secrets. In this way, multinational companies operating within both the United States and Europe can maintain a single system for trade secret asset management and obtain legal protections for their trade secret information in both territories.

Summation

The automation of trade secret asset management, once considered impossible, has become practical. The benefits include reduced costs, increased security, and a greater likelihood of succeeding on the merits if litigation becomes necessary. Employee entry, made possible by easy-to-understand methods of categorization and scoring, achieve the desired benefits while containing risk by leaving trade secrets where they are. Blockchaining hash codes of the information about trade secrets creates an unassailable historical record. As automated systems become easier to use and more available, the reasonable measures required by trade secret information owners will likely evolve to include automated trade secret asset management.

Chapter 20
The Trade Secret Examiner® System

Executive Summary

A trade secret asset management system that meets all the requirements of Chapter 19 is now commercially available. The Trade Secret Examiner® system was designed to implement the authors' strategies for trade secret protection. This summary of its capabilities is included here for interested readers. A complete feature list is contained in Appendix G. More information on the Trade Secret Examiner® system is available from the Trade Secret Office, Inc.

Overall Approach

The goal of the Trade Secret Examiner® system is to create and maintain an inventory of trade secrets within the company. It does this through the means of a database containing information about the trade secrets.

The best analogy for the operation of the Trade Secret Examiner® system is a library card catalog. The library card catalog entry for a book in the library contains the name of the book, the name of the author, the year the book was published, the topic of the book, and where the book is located in the company. Importantly, it does not contain the book itself, or any of the text of the book.

Similarly, the Trade Secret Examiner® system maintains

entries for the trade secrets within the company. Such an entry contains the name of the trade secret, the creator of the trade secret, the date the trade secret was created, the SFP of the trade secret, and where the trade secret is located within the company. The Trade Secret Examiner® system does not contain the trade secret itself, just as the library card catalog does not contain the book.

The Trade Secret Examiner® system is a browser-based web application that can run on a single machine, on a server on the company's intranet, or on a cloud-based server.

The Trade Secret Worksheet

Trade secrets are entered into the system by employees who already know the trade secret information. The trade secret information does not need to be exposed to any additional employees to enter or manage trade secrets within the system.

Trade secrets are entered on the Trade Secret Worksheet. The worksheet is similar to any web form, with drop-down boxes, free-form fields, and checkboxes. The employee enters a name for the trade secret, the date the trade secret when the trade secret was created, selects the Subject, Format, and Product for the trade secret, and applies a one-to-five score for each of the six factors. The employee also selects keywords that apply to the trade secret from a list, or enters new keywords that apply. The employee also specifies the physical location in which the trade secret exists, as well as the organizational group in which it is known.

The Trade Secret Worksheet, in addition to traditional browser access, can be used from an iPad or a smartphone, allowing the entry of trade secrets from the shop floor as well as the office. The Trade Secret Worksheet is accessible within all privilege levels within the Trade Secret Examiner® system. Other operations require Analyst, Reviewer, Litigator, or Admin privileges.

The trade secrets entered by the employees through the Trade Secret Worksheet become Potential Trade Secrets within the Trade Secret Examiner® system.

Acceptance of Potential Trade Secrets

Once a Potential Trade Secret has been entered by an employee who already knows the trade secret information, management of the metadata entered by the employee is the role of the intellectual property group charged with trade secret management. This group should operate under the direction of an attorney so attorney client privilege applies to its operation of the Trade Secret Examiner® system. The Potential Trade Secret is reviewed, and a decision is made as to whether to include it in the Trade Secret Inventory.

At this time, the information is also reviewed for consistency based on the metadata alone. A Potential Trade Secret with a name of "Method of machining engine cylinder heads," for example, should not be assigned a Subject of Accounting, or a Format of Spreadsheet, or a Product of Tires.

The intellectual property group can review the values for the Six Factors, based on the Potential Trade Secret's name, without actually knowing what the machining method is. They can also review the keywords, changing keywords or assigning additional keywords.

If the Potential Trade Secret is marked Accepted by the intellectual property group, it joins the inventory of the company's Trade Secrets.

Database Permanence

The database design of the Trade Secret Examiner® system is different from most databases with which the reader may be familiar, in that it retains all information forever.

If you think of a contact database for example, once you delete a contact, the contact is gone. We are all familiar with the message "Are you sure you want to delete this contact? This deletion cannot be undone." If you change the phone number, address, or employer of a contact, the previous information is gone.

Not so with the Trade Secret Examiner® system. Any information entered is retained forever. When information is changed, the old version of the information, and the date range during which it was effective, is retained forever. If it were a contact database, you would be able to cycle through the past phone numbers, addresses, and employers of any contact in the database, and deleted contacts would go into a separate location from which they could be retrieved.

Due to this structure, the information that applied to any Trade Secret in the Trade Secret Examiner® system at any time in the past can be recreated as of that date. This is particularly helpful when preparing documents for trial.

Redundancy Elimination

There are many benefits to having the employees who already know the trade secret information enter the trade secrets into the Trade Secret Examiner® system. One disadvantage is that multiple employees may enter the same trade secret, producing multiple Potential Trade Secrets that actually refer to the same underlying trade secret information.

The Trade Secret Examiner® system has a redundancy elimination feature that finds and then combines multiple redundant Potential Trade Secrets into one Trade Secret in the inventory. The multiple scores for each of the Six Factors are averaged from the multiple redundant entries to refine the values based on the combined judgment of the employees who entered it.

Trade Secrets Involved In Litigation

When litigation is necessary, the Trade Secret Examiner® system can be used to gain insight as to which trade secrets to assert at trial. Selecting trade secrets by Location, Group, and then ranking them by Defendability Factor produces a list of the trade secrets to which the defendant had access ranked by the company's ability to prove them in court. Other selections, based on such things as the economic value to the company, are also available.

The Trade Secret Examiner® system also tracks which trade secrets have been asserted in multiple litigation matters, and provides tools for assigning a litigation code to each trade secret asserted in each litigation, so it can be referred to at trial without actually naming the trade secret.

Security Analysis

The Trade Secret Examiner® system provides security analysis tools to ensure that security for trade secrets is appropriate to their economic value and defendability factor. This prevents both the alpha error of insufficient security for important trade secrets, as well as the beta error of spending too much money providing excessive security for lesser trade secrets.

Non-Disclosure and Employee Agreements

Executed non-disclosure agreements and employee agreements are often lost or misplaced, preventing their use at trial. Absent this evidence, plaintiff in a trade secrets case may have a hard time proving there was any burden of confidentiality assumed, or that an assignment of rights in trade secrets was made.

The Trade Secret Examiner® system provides means to store the executed agreements within the database, ensuring against them being misplaced. Further, the executed agreements are verified with the same blockchain technology as the rest of the database, preventing any challenge as to provenance in court.

Blockchaining

The Trade Secret Examiner® system uses blockchaining to prove the existence of data by a date certain, for all the data in the database. The current data, the historical data, the executed non-disclosure agreements, the executed employee agreements, are all blockchained as of the date they were added to or changed within the database. There can be no argument that the materials produced were falsified or created after the fact.

The Trade Secret Examiner® system uses a unique two-stage blockchain to limit communications from the Trade Secret Examiner® server to the blockchain host, while maintaining the full benefit of blockchain technology.

Summation

A full-featured trade secret asset management system is now commercially available. The Trade Secret Examiner® system implements the authors' strategies for trade secret protection, including all the required features as described in Chapter 19, including European compliance.

APPENDICES

Appendix A
Trade Secrets and the Law

Appendix A.1

UNIFORM TRADE SECRETS ACT WITH 1985 AMENDMENTS

SECTION 1. DEFINITIONS. As used in this [Act], unless the context requires otherwise:

(1) "Improper means" includes theft, bribery, misrepresentation, breach or inducement of a breach of a duty to maintain secrecy, or espionage through electronic or other means;

(2) "Misappropriation" means:

(i) acquisition of a trade secret of another by a person who knows or has reason to know that the trade secret was acquired by improper means; or

(ii) disclosure or use of a trade secret of another without express or implied consent by a person who

(A) used improper means to acquire knowledge of the trade secret; or

(B) at the time of disclosure or use, knew or had reason to know that his knowledge of the trade secret was

(I) derived from or through a person who had utilized improper means to acquire it;

(II) acquired under circumstances giving rise to a duty to maintain its secrecy or limit its use; or

(III) derived from or through a person who owed a duty to the person seeking relief to maintain its secrecy or limit its use; or

(C) before a material change of his [or her] position, knew or had reason to know that it was a trade secret and that knowledge of it had been acquired by accident or mistake.

(3) "Person" means a natural person, corporation, business trust, estate, trust, partnership, association, joint venture, government, governmental subdivision or agency, or any other legal or commercial entity.

(4) "Trade secret" means information, including a formula, pattern, compilation, program, device, method, technique, or process, that:

(i) derives independent economic value, actual or potential, from not being generally known to, and not being readily ascertainable by proper means by, other persons who can obtain economic value from its disclosure or use, and

(ii) is the subject of efforts that are reasonable under the circumstances to maintain its secrecy.

SECTION 2. INJUNCTIVE RELIEF.

(a) Actual or threatened misappropriation may be enjoined. Upon application to the court, an injunction shall be terminated when the trade secret has ceased to exist, but the injunction may be continued for an additional reasonable period of time in order to eliminate commercial advantage that otherwise would be derived from the misappropriation.

(b) In exceptional circumstances, an injunction may condition future use upon payment of a reasonable royalty for no longer than the period of time for which use could have been prohibited. Exceptional circumstances include, but are not limited to, a material and prejudicial change of position prior to acquiring knowledge or reason to know of misappropriation that renders a prohibitive injunction inequitable.

(c) In appropriate circumstances, affirmative acts to protect a trade secret may be compelled by court order.

SECTION 3. DAMAGES.

(a) Except to the extent that a material and prejudicial change of position prior to acquiring knowledge or reason to know of misappropriation renders a monetary recovery inequitable, a complainant is entitled to recover damages for misappropriation. Damages can include both the actual loss caused by misappropriation and the unjust enrichment caused by misappropriation that is not taken into account in computing actual loss. In lieu of damages measured by any other methods, the damages caused by misappropriation may be measured by imposition of liability for a reasonable royalty for a misappropriator's unauthorized disclosure or use of a trade secret.

(b) If willful and malicious misappropriation exists, the court may award exemplary damages in an amount not exceeding twice any award made under subsection (a).

SECTION 4. ATTORNEY'S FEES.

If (i) a claim of misappropriation is made in bad faith, (ii) a motion to terminate an injunction is made or resisted in bad faith, or (iii) willful and malicious misappropriation exists, the court may award reasonable attorney's fees to the prevailing party.

SECTION 5. PRESERVATION OF SECRECY.

In an action under this [Act], a court shall preserve the secrecy of an alleged trade secret by reasonable means, which may include granting protective orders in connection with discovery proceedings, holding in-camera hearings, sealing the records of the action, and ordering any person involved in the litigation not to disclose an alleged trade secret without prior court approval.

SECTION 6. STATUTE OF LIMITATIONS.

An action for misappropriation must be brought within 3 years after the misappropriation is discovered or by the exercise of reasonable diligence should have been discovered. For the purposes of this section, a continuing misappropriation constitutes a single claim.

SECTION 7. EFFECT ON OTHER LAW.

(a) Except as provided in subsection (b), this [Act] displaces conflicting tort, restitutionary, and other law of this State providing civil remedies for misappropriation of a trade secret.

(b) This [Act] does not affect:

(1) contractual remedies, whether or not based upon misappropriation of a trade secret;

(2) other civil remedies that are not based upon misappropriation of a trade secret~~.~~ ; or

(3) criminal remedies, whether or not based upon misappropriation of a trade secret.

SECTION 8. UNIFORMITY OF APPLICATION AND CONSTRUCTION.

This [Act] shall be applied and construed to effectuate its general purpose to make uniform the law with respect to the subject of this [Act] among states enacting it.

SECTION 9. SHORT TITLE.

This [Act] may be cited as the Uniform Trade Secrets Act.

SECTION 10. SEVERABILITY.

If any provision of this [Act] or its application to any person or circumstances is held invalid, the invalidity does not affect other provisions or applications of the [Act] which can be given effect without the invalid provision or application, and to this end the provisions of this [Act] are severable.

SECTION 11. TIME OF TAKING EFFECT.

This [Act] takes effect on ________________, and does not apply to misappropriation occurring prior to the effective date. With respect to a continuing misappropriation that began prior to the effective date, the [Act] also does not apply to the continuing misappropriation that occurs after the effective date.

Appendix A.2

Section 757 comment b of the Restatement (First) of Torts (1939)

Definition of trade secret. A trade secret may consist of any formula, pattern, device or compilation of information which is used on one's business, and which gives him an opportunity to obtain an advantage over competitors who do not know or use it. It may be a formula for a chemical compound, a process of manufacturing, treating or preserving materials, a pattern for a machine or other device, or a list of customers. It differs from other secret information in a business in that it is not simply information as to single or ephemeral events in the conduct of the business, as, for example, the amount of other terms of a secret bid for a contract or the salary of certain employees, or the security investments made or contemplated, or the date fixed for the announcement of a new policy or for bringing out a new model or the like. A trade secret is a process or device for continuous use in the operations of the business. Generally it relates to the production of goods, as, for example, a machine or formula for the production of an article. It may, however, relate to the sale of goods or to other operations in the business, such as a code for determining discounts, rebates or other concessions in a price list or catalogue, or a list of specialized customers, or a method of bookkeeping or other office management.

Secrecy. The subject matter of a trade secret must be secret. Matters of public knowledge or of general knowledge in an industry cannot be appropriated by one as his secret. Matters which are completely disclosed by the goods which one markets cannot be his secret. Substantially, a trade secret is known only in the particular business in which it is used. It is not requisite that only the proprietor of the business know it. He may, without losing his protection, communicate it to employees involved in its use. He may likewise communicate it to others pledged to secrecy. Others may also know of it independently, as, for example, when they have discovered the process or formula by independent invention and are keeping it secret. Nevertheless, a substantial element of secrecy must exist, so that, except by the use of improper means, there would be difficulty in acquiring the information. An exact definition of a trade secret is not possible. Some factors to be considered in determining whether given information is one's trade secret are:

(1) The extent to which the information is known outside of his business;

(2) The extent to which it is known by employees and others involved in his business;

(3) The extent of measures taken by him to guard the secrecy of the information;

(4) The value of the information to him and his competitors;

(5) The amount of effort or money expended by him in developing the information;

(6) The ease or difficulty with which the information could be properly acquired or duplicated by others.

Novelty and prior art. A trade secret may be a device or process which is patentable; but it need not be that. It may be a device or process which is clearly anticipated in the prior art or one which is merely a mechanical improvement that a good mechanic can make. Novelty and invention are not requisite for a trade secret as they are for patentability. These requirements are essential to patentability because a patent protects against unlicensed use of the patented device or process even by one who discovers it properly through independent research. The patent monopoly is a reward to the inventor. But such is not the case with a trade secret. Its protection is not based on a policy of rewarding or otherwise encouraging the development of secret processes or devices. The protection is merely against breach of faith and reprehensible means of learning another's secret. For this limited protection it is not appropriate to require also the kind of novelty and invention which is a requisite of patentability. The nature of the secret is, however, an important factor in determining the kind of relief that is appropriate against one who is subject to the liability under the rule stated in this section. Thus, if the secret consists of a device or process which is a novel invention, one who acquires the secret wrongfully is ordinarily enjoined from further use of it and is required to account for the profits derived from his past use. If, on the other hand, the secret consists of mechanical improvements that a good mechanic can make without resort to the secret, the wrongdoer's liability may be limited to damages, and an injunction against future use of the improvements made with the aid of the secret may be inappropriate.

Appendix A.3

THE ECONOMIC ESPIONAGE ACT OF 1996

United States Code, Section 18

§ 1831. Economic espionage

(a) In General.— Whoever, intending or knowing that the offense will benefit any foreign government, foreign instrumentality, or foreign agent, knowingly—

(1) steals, or without authorization appropriates, takes, carries away, or conceals, or by fraud, artifice, or deception obtains a trade secret;

(2) without authorization copies, duplicates, sketches, draws, photographs, downloads, uploads, alters, destroys, photocopies, replicates, transmits, delivers, sends, mails, communicates, or conveys a trade secret;

(3) receives, buys, or possesses a trade secret, knowing the same to have been stolen or appropriated, obtained, or converted without authorization;

(4) attempts to commit any offense described in any of paragraphs (1) through (3); or

(5) conspires with one or more other persons to commit any offense described in any of paragraphs (1) through (4), and one or more of such persons do any act to effect the object of conspiracy,

shall, except as provided in subsection (b), be fined not more than $500,000 or imprisoned not more than 15 years, or both.

(b) ORGANIZATIONS.- Any organization that commits any offense described in subsection (a) shall be fined not more than $10,000,000.

§ 1832. Theft of trade secrets

(a) Whoever, with intent to convert a trade secret, that is related to or included in a product that is produced for or placed in interstate or foreign commerce, to the economic benefit of anyone other than the owner thereof, and intending or knowing that the offense will, injure any owner of that trade secret, knowingly—

(1) steals, or without authorization appropriates, takes, carries away, or conceals, or by fraud, artifice, or deception obtains such information;

(2) without authorization copies, duplicates, sketches, draws, photographs, downloads, uploads, alters, destroys, photocopies, replicates, transmits, delivers, sends, mails, communicates, or conveys such information;

(3) receives, buys, or possesses such information, knowing the same to have been stolen or appropriated, obtained, or converted without authorization;

(4) attempts to commit any offense described in paragraphs (1) through (3); or

(5) conspires with one or more other persons to commit any offense described in paragraphs (1) through (3), and one or more of such persons do any act to effect the object of the conspiracy,

shall, except as provided in subsection (b), be fined under this title or imprisoned not more than 10 years, or both.

(b) Any organization that commits any offense described in subsection (a) shall be fined not more than $5,000,000.

§ 1833. Exceptions to prohibitions

This chapter does not prohibit—

(1) any otherwise lawful activity conducted by a governmental entity of the United States, a State, or a political subdivision of a State; or

(2) the reporting of a suspected violation of law to any governmental entity of the United States, a State, or a political subdivision of a State, if such entity has lawful authority with respect to that violation.

§ 1834. Criminal forfeiture

(a) The court, in imposing sentence on a person for a violation of this chapter, shall order, in addition to any other sentence imposed, that the person forfeit to the United States—

(1) any property constituting, or derived from, any proceeds the person obtained, directly or indirectly, as the result of such violation; and

(2) any of the person's property used, or intended to be used, in any manner or part, to commit or facilitate the commission of such violation, if the court in its discretion so determines, taking into consideration the nature, scope, and proportionality of the use of the property in the offense.

(b) Property subject to forfeiture under this section, any seizure and disposition thereof, and any administrative or judicial proceeding in relation thereto, shall be governed by section 413 of the Comprehensive Drug Abuse Prevention and Control Act of 1970 (21 U.S.C. 853), except for subsections (d) and (j) of such section, which shall not apply to forfeitures under this section.

§ 1835. Orders to preserve confidentiality

In any prosecution or other proceeding under this chapter, the court shall enter such orders and take such other action as may be necessary and appropriate to preserve the confidentiality of trade secrets, consistent with the requirements of the Federal Rules of Criminal and Civil Procedure, the Federal Rules of Evidence, and all other applicable laws. An interlocutory appeal by the United States shall lie from a decision or order of a district court authorizing or directing the disclosure of any trade secret.

§ 1836. Civil proceedings to enjoin violations

(a) The Attorney General may, in a civil action, obtain appropriate injunctive relief against any violation of this section.

(b) The district courts of the United States shall have exclusive original jurisdiction of civil actions under this subsection.

§ 1837. Applicability to conduct outside the United States

This chapter also applies to conduct occurring outside the United States if—

(1) the offender is a natural person who is a citizen or permanent resident alien of the United States, or an organization organized under the laws of the United States or a State or political subdivision thereof; or

(2) an act in furtherance of the offense was committed in the United States.

§ 1838. Construction with other laws

This chapter shall not be construed to preempt or displace any other remedies, whether civil or criminal, provided by United States Federal, State, commonwealth, possession, or territory law for the misappropriation of a trade secret, or to affect the otherwise lawful disclosure of information by any Government employee under section 552 of title 5 (commonly known as the Freedom of Information Act).

§ 1839. Definitions

As used in this chapter

(1) the term 'foreign instrumentality' means any agency, bureau, ministry, component, institution, association, or any legal, commercial, or business organization, corporation, firm, or entity that is substantially owned; controlled, sponsored, commanded, managed, or dominated by a foreign government;

(2) the term 'foreign agent' means any officer, employee, proxy, servant, delegate, or representative of a foreign government;

(3) the term 'trade secret' means all forms and types of financial, business, scientific, technical, economic, or engineering information, including patterns, plans, compilations, program devices, formulas, designs, prototypes, methods, techniques, processes, procedures, programs, or codes, whether tangible or intangible, and whether or how stored, compiled, or memorialized physically, electronically, graphically, photographically, or in writing if—

(A) the owner thereof has taken reasonable measures to keep such information secret; and

(B) the information derives independent economic value, actual or potential, from not being generally known to, and not being readily ascertainable through proper means by, the public; and
(4) the term 'owner', with respect to a trade secret, means the person or entity in whom or in which rightful legal or equitable title to , or license in, the trade secret is reposed.

Appendix A.4
THE DEFEND TRADE SECRETS ACT OF 2016

Public Law 114-153
114th Congress

An Act

To amend chapter 90 of title 18, United States Code, to provide Federal jurisdiction for the theft of trade secrets, and for other purposes.

Be it enacted by the Senate and House of Representatives of the United States of America in Congress assembled,

SECTION 1. SHORT TITLE.

This Act may be cited as the ``Defend Trade Secrets Act of 2016''.

SEC. 2. FEDERAL JURISDICTION FOR THEFT OF TRADE SECRETS.

(a) In General.--Section 1836 of title 18, United States Code, is amended by striking subsection (b) and inserting the following:

``(b) Private Civil Actions.--

``(1) In general.--An owner of a trade secret that is misappropriated may bring a civil action under this subsection if the trade secret is related to a product or service used in, or intended for use in, interstate or foreign commerce.

``(2) Civil seizure.--

``(A) In general.--

``(i) Application.--Based on an affidavit or verified complaint satisfying the requirements of this paragraph, the court may, upon ex parte application but only in extraordinary circumstances, issue an order providing for the seizure of property necessary to prevent the propagation or dissemination of the trade secret that is the subject of the action.

``(ii) Requirements for issuing order.--The court may not grant an application under clause (i) unless the court finds that it clearly appears from specific facts that--

``(I) an order issued pursuant to Rule 65 of the Federal Rules of Civil Procedure or another form of equitable relief would be inadequate to achieve the purpose of this paragraph because the party to which the order would be issued would evade, avoid, or otherwise not comply with such an order;

``(II) an immediate and irreparable injury will occur if such seizure is not ordered;

``(III) the harm to the applicant of denying the application outweighs the harm to the legitimate interests of the person against whom seizure would be ordered of granting the application and substantially outweighs the harm to any third parties who may be harmed by such seizure;

``(IV) the applicant is likely to succeed in showing that--

``(aa) the information is a trade secret; and

``(bb) the person against whom seizure would be ordered--

``(AA) misappropriated the trade secret of the applicant by improper means; or

``(BB) conspired to use improper means to

misappropriate the trade secret of the applicant;

``(V) the person against whom seizure would be ordered has actual possession of--

``(aa) the trade secret; and

``(bb) any property to be seized;

``(VI) the application describes with reasonable particularity the matter to be seized and, to the extent reasonable under the circumstances, identifies the location where the matter is to be seized;

``(VII) the person against whom seizure would be ordered, or persons acting in concert with such person, would destroy, move, hide, or otherwise make such matter inaccessible to the court, if the applicant were to proceed on notice to such person; and

``(VIII) the applicant has not publicized the requested seizure.

``(B) Elements of order.--If an order is issued under subparagraph (A), it shall--

``(i) set forth findings of fact and conclusions of law required for the order;

``(ii) provide for the narrowest seizure of property necessary to achieve the purpose of this paragraph and direct that the seizure be conducted in a manner that minimizes any interruption of the business operations of third parties and, to the extent possible, does not interrupt the legitimate business operations of the person accused of misappropriating the trade secret;

``(iii)(I) be accompanied by an order protecting the seized property from disclosure by prohibiting access by the applicant or the person against whom the order is directed, and prohibiting any copies, in whole or in part, of the seized property, to prevent undue damage to the party against whom the order has issued or others, until such parties have an opportunity to be heard in court; and

``(II) provide that if access is granted by the court to the applicant or the person against whom the order is directed, the access shall be consistent with subparagraph (D);

``(iv) provide guidance to the law enforcement officials executing the seizure that clearly delineates the scope of the authority of the officials, including--

``(I) the hours during which the seizure may be executed; and

``(II) whether force may be used to access locked areas;

``(v) set a date for a hearing described in subparagraph (F) at the earliest possible time, and not later than 7 days after the order has issued, unless the party against whom the order is directed and others harmed by the order consent to another date for the hearing, except that a party against whom the order has issued or any person harmed by the order may move the court at any time to dissolve or modify the order after giving notice to the applicant who obtained the order; and

``(vi) require the person obtaining the order to provide the security determined adequate by the court for the payment of the damages that any person may be entitled to recover as a result of a wrongful or excessive seizure or wrongful or excessive attempted seizure under this paragraph.

``(C) Protection from publicity.--The court shall take appropriate action to protect the person against whom an order under this paragraph is directed from publicity, by or at the behest of the person obtaining the order, about such order and any seizure under such order.

``(D) Materials in custody of court.--

``(i) In general.--Any materials seized under this paragraph shall be taken into the custody of the court. The court shall secure the seized material from physical and electronic access during the seizure and while in the custody of the court.

``(ii) Storage medium.--If the seized material includes a storage medium, or if the seized material is stored on a storage medium, the court shall prohibit the medium from being connected to a network or the Internet without the consent of both parties, until the hearing required under subparagraph (B)(v) and described in subparagraph (F).

``(iii) Protection of confidentiality.--The court shall take appropriate measures to protect the confidentiality of seized materials that are unrelated to the trade secret information ordered seized pursuant to this paragraph unless the person against whom the order is entered consents to disclosure of the material.

``(iv) Appointment of special master.--The court may appoint a special master to locate and isolate all misappropriated trade secret information and to facilitate the return of unrelated property and data to the person from whom the property was seized. The special master appointed by the court shall agree to be bound by a non-disclosure agreement approved by the court.

``(E) Service of order.--The court shall order that service of a copy of the order under this paragraph, and the submissions of the applicant to obtain the order, shall be made by a Federal law enforcement officer who, upon making service, shall carry out the seizure under the order. The court may allow State or local law enforcement officials to participate, but may not permit the applicant or any agent of the applicant to participate in the seizure. At the request of law enforcement officials, the court may allow a technical expert who is unaffiliated with the applicant and who is bound by a court-approved non-disclosure agreement to participate in the seizure if the court determines that the participation of the expert will aid the efficient execution of and minimize the burden of the seizure.

``(F) Seizure hearing.--

``(i) Date.--A court that issues a seizure order shall hold a hearing on the date set by the court under subparagraph (B)(v).

``(ii) Burden of proof.--At a hearing held under this subparagraph, the party who obtained the order under subparagraph (A) shall have the burden to prove the facts supporting the findings of fact and conclusions of law necessary to support the order. If the party fails to meet that burden, the seizure order shall be dissolved or modified appropriately.

``(iii) Dissolution or modification of order.--A party against whom the order has been issued or any person harmed by the order may move the court at any time to dissolve or modify the order after giving notice to the party who obtained the order.

``(iv) Discovery time limits.--The court may make such orders modifying the time limits for discovery under the Federal Rules of Civil Procedure as may be necessary to prevent the frustration of the purposes of a hearing under this subparagraph.

``(G) Action for damage caused by wrongful

seizure.--A person who suffers damage by reason of a wrongful or excessive seizure under this paragraph has a cause of action against the applicant for the order under which such seizure was made, and shall be entitled to the same relief as is provided under section 34(d)(11) of the Trademark Act of 1946 (15 U.S.C. 1116(d)(11)). The security posted with the court under subparagraph (B)(vi) shall not limit the recovery of third parties for damages.

``(H) Motion for encryption.--A party or a person who claims to have an interest in the subject matter seized may make a motion at any time, which may be heard ex parte, to encrypt any material seized or to be seized under this paragraph that is stored on a storage medium. The motion shall include, when possible, the desired encryption method.

``(3) Remedies.--In a civil action brought under this subsection with respect to the misappropriation of a trade secret, a court may--

``(A) grant an injunction--

``(i) to prevent any actual or threatened misappropriation described in paragraph (1) on such terms as the court deems reasonable, provided the order does not--

``(I) prevent a person from entering into an employment relationship, and that conditions placed on such employment shall be based on evidence of threatened misappropriation and not merely on the information the person knows; or

``(II) otherwise conflict with an applicable State law prohibiting restraints on the practice of a lawful profession, trade, or business;

``(ii) if determined appropriate by the court, requiring affirmative actions to be taken to

protect the trade secret; and
``(iii) in exceptional circumstances that render an injunction inequitable, that conditions future use of the trade secret upon payment of a reasonable royalty for no longer than the period of time for which such use could have been prohibited;
``(B) award--
``(i)(I) damages for actual loss caused by the misappropriation of the trade secret; and
``(II) damages for any unjust enrichment caused by the misappropriation of the trade secret that is not addressed in computing damages for actual loss; or
``(ii) in lieu of damages measured by any other methods, the damages caused by the misappropriation measured by imposition of liability for a reasonable royalty for the misappropriator's unauthorized disclosure or use of the trade secret;
``(C) if the trade secret is willfully and maliciously misappropriated, award exemplary damages in an amount not more than 2 times the amount of the damages awarded under subparagraph (B); and
``(D) if a claim of the misappropriation is made in bad faith, which may be established by circumstantial evidence, a motion to terminate an injunction is made or opposed in bad faith, or the trade secret was willfully and maliciously misappropriated, award reasonable attorney's fees to the prevailing party.
``(c) Jurisdiction.--The district courts of the United States shall have original jurisdiction of civil actions brought under this section.
``(d) Period of Limitations.--A civil action under subsection (b) may not be commenced later than 3 years after the date on which the misappropriation with respect to which the action would relate is discovered or by the exercise of reasonable diligence should have been discovered. For purposes of this subsection, a continuing

misappropriation constitutes a single claim of misappropriation.".

(b) Definitions.--Section 1839 of title 18, United States Code, is amended--

(1) in paragraph (3)--

(A) in subparagraph (B), by striking ``the public'' and inserting ``another person who can obtain economic value from the disclosure or use of the information''; and

(B) by striking ``and'' at the end;

(2) in paragraph (4), by striking the period at the end and inserting a semicolon; and

(3) by adding at the end the following:

``(5) the term `misappropriation' means--

``(A) acquisition of a trade secret of another by a person who knows or has reason to know that the trade secret was acquired by improper means; or

``(B) disclosure or use of a trade secret of another without express or implied consent by a person who--

``(i) used improper means to acquire knowledge of the trade secret;

``(ii) at the time of disclosure or use, knew or had reason to know that the knowledge of the trade secret was--

``(I) derived from or through a person who had used improper means to acquire the trade secret;

``(II) acquired under circumstances giving rise to a duty to maintain the secrecy of the trade secret or limit the use of the trade secret; or

``(III) derived from or through a person who owed a duty to the person seeking relief to maintain the secrecy of the trade secret or limit the use of the trade secret; or

``(iii) before a material change of the position of the person, knew or had reason to know that--

``(I) the trade secret was a trade secret; and

``(II) knowledge of the trade secret had been acquired by accident or mistake;

``(6) the term `improper means'--

``(A) includes theft, bribery, misrepresentation, breach or inducement of a breach of a duty to maintain secrecy, or espionage through electronic or other means; and

``(B) does not include reverse engineering, independent derivation, or any other lawful means of acquisition; and

``(7) the term `Trademark Act of 1946' means the Act entitled `An Act to provide for the registration and protection of trademarks used in commerce, to carry out the provisions of certain international conventions, and for other purposes, approved July 5, 1946 (15 U.S.C. 1051 et seq.) (commonly referred to as the ``Trademark Act of 1946'' or the ``Lanham Act'')'.''.

(c) Exceptions to Prohibition.--Section 1833 of title 18, United States Code, is amended, in the matter preceding paragraph (1), by inserting ``or create a private right of action for'' after ``prohibit''.

(d) Conforming Amendments.--

(1) The section heading for section 1836 of title 18, United States Code, is amended to read as follows:

``Sec. 1836. Civil proceedings''.

(2) The table of sections for chapter 90 of title 18, United States Code, is amended by striking the item relating to section 1836 and inserting the following:

``1836. Civil proceedings.''.

(e) Effective Date.--The amendments made by this section shall apply with respect to any misappropriation of a trade secret (as defined in section 1839 of title 18, United States Code, as amended by this section) for which any act occurs on or after the date of the enactment of this Act.

(f) Rule of Construction.--Nothing in the amendments made by this section shall be construed to modify the rule of construction under section 1838 of title 18, United States Code, or to preempt any other provision of law.

(g) Applicability to Other Laws.--This section and the amendments made by this section shall not be construed to be a law pertaining to intellectual property for purposes of any other Act of Congress.

SEC. 3. TRADE SECRET THEFT ENFORCEMENT.

(a) In General.--Chapter 90 of title 18, United States Code, is amended--

(1) in section 1832(b), by striking ``$5,000,000'' and inserting ``the greater of $5,000,000 or 3 times the value of the stolen trade secret to the organization, including expenses for research and design and other costs of reproducing the trade secret that the organization has thereby avoided''; and

(2) in section 1835--

(A) by striking ``In any prosecution'' and inserting the following:

``(a) In General.--In any prosecution''; and

(B) by adding at the end the following:

``(b) Rights of Trade Secret Owners.--The court may not authorize or direct the disclosure of any information the owner asserts to be a trade secret unless the court allows the owner the opportunity to file a submission under seal that describes the interest of the owner in keeping the information confidential. No submission under seal made under this subsection may be used in a prosecution under this chapter for any purpose other than those set forth in this section, or otherwise required by law. The provision of information relating to a trade secret to the United States or the court in connection with a prosecution under this chapter shall not constitute a waiver of trade secret protection, and the disclosure of information relating to a trade secret in connection with a prosecution under this chapter shall not constitute a waiver of trade secret protection unless the trade secret owner expressly consents to such waiver.''.

(b) RICO Predicate Offenses.--Section 1961(1) of title 18, United States Code, is amended by inserting ``sections 1831 and 1832 (relating to economic espionage and theft of trade secrets)," before ``section 1951".

SEC. 4. REPORT ON THEFT OF TRADE SECRETS OCCURRING ABROAD.

(a) Definitions.--In this section:

(1) Director.--The term ``Director" means the Under Secretary of Commerce for Intellectual Property and Director of the United States Patent and Trademark Office.

(2) Foreign instrumentality, etc.--The terms ``foreign instrumentality", ``foreign agent", and ``trade secret" have the meanings given those terms in section 1839 of title 18, United States Code.

(3) State.--The term ``State" includes the District of Columbia and any commonwealth, territory, or possession of the United States.

(4) United states company.--The term ``United States company" means an organization organized under the laws of the United States or a State or political subdivision thereof.

(b) Reports.--Not later than 1 year after the date of enactment of this Act, and biannually thereafter, the Attorney General, in consultation with the Intellectual Property Enforcement Coordinator, the Director, and the heads of other appropriate agencies, shall submit to the Committees on the Judiciary of the House of Representatives and the Senate, and make publicly available on the Web site of the Department of Justice and disseminate to the public through such other means as the Attorney General may identify, a report on the following:

(1) The scope and breadth of the theft of the trade secrets of United States companies occurring outside of the United States.

(2) The extent to which theft of trade secrets occurring outside of the United States is sponsored by foreign governments, foreign instrumentalities, or foreign agents.

(3) The threat posed by theft of trade secrets occurring outside of the United States.

(4) The ability and limitations of trade secret owners to prevent the misappropriation of trade secrets outside of the United States, to enforce any judgment against foreign entities for theft of trade secrets, and to prevent imports based on theft of trade secrets overseas.

(5) A breakdown of the trade secret protections afforded United States companies by each country that is a trading partner of the United States and enforcement efforts available and undertaken in each such country, including a list identifying specific countries where trade secret theft, laws, or enforcement is a significant problem for United States companies.

(6) Instances of the Federal Government working with foreign countries to investigate, arrest, and prosecute entities and individuals involved in the theft of trade secrets outside of the United States.

(7) Specific progress made under trade agreements and treaties, including any new remedies enacted by foreign countries, to protect against theft of trade secrets of United States companies outside of the United States.

(8) Recommendations of legislative and executive branch actions that may be undertaken to--

(A) reduce the threat of and economic impact caused by the theft of the trade secrets of United States companies occurring outside of the United States;

(B) educate United States companies regarding the threats to their trade secrets when taken outside of the United States;

(C) provide assistance to United States companies to reduce the risk of loss of their trade secrets when taken outside of the United States; and

(D) provide a mechanism for United States companies to confidentially or anonymously report the theft of trade secrets occurring outside of the United States.

SEC. 5. SENSE OF CONGRESS.

It is the sense of Congress that--

(1) trade secret theft occurs in the United States and around the world;

(2) trade secret theft, wherever it occurs, harms the companies that own the trade secrets and the employees of the companies;

(3) chapter 90 of title 18, United States Code (commonly known as the ``Economic Espionage Act of 1996"), applies broadly to protect trade secrets from theft; and

(4) it is important when seizing information to balance the need to prevent or remedy misappropriation with the need to avoid interrupting the--

(A) business of third parties; and

(B) legitimate interests of the party accused of wrongdoing.

SEC. 6. BEST PRACTICES.

(a) In General.--Not later than 2 years after the date of enactment of this Act, the Federal Judicial Center, using existing resources, shall develop recommended best practices for--

(1) the seizure of information and media storing the information; and

(2) the securing of the information and media once seized.

(b) Updates.--The Federal Judicial Center shall update the recommended best practices developed under subsection (a) from time to time.

(c) Congressional Submissions.--The Federal Judicial Center shall provide a copy of the recommendations developed under subsection (a), and any updates made under subsection (b), to the--

(1) Committee on the Judiciary of the Senate; and

(2) Committee on the Judiciary of the House of Representatives.

SEC. 7. IMMUNITY FROM LIABILITY FOR CONFIDENTIAL

DISCLOSURE OF A TRADE SECRET TO THE GOVERNMENT OR IN A COURT FILING.

(a) Amendment.--Section 1833 of title 18, United States Code, is amended--

(1) by striking ``This chapter'' and inserting ``(a) In General.--This chapter'';

(2) in subsection (a)(2), as designated by paragraph (1), by striking ``the reporting of a suspected violation of law to any governmental entity of the United States, a State, or a political subdivision of a State, if such entity has lawful authority with respect to that violation'' and inserting ``the disclosure of a trade secret in accordance with subsection (b)''; and

(3) by adding at the end the following:

``(b) Immunity From Liability for Confidential Disclosure of a Trade Secret to the Government or in a Court Filing.--

``(1) Immunity.--An individual shall not be held criminally or civilly liable under any Federal or State trade secret law for the disclosure of a trade secret that--

``(A) is made--

``(i) in confidence to a Federal, State, or local government official, either directly or indirectly, or to an attorney; and

``(ii) solely for the purpose of reporting or investigating a suspected violation of law; or

``(B) is made in a complaint or other document filed in a lawsuit or other proceeding, if such filing is made under seal.

``(2) Use of trade secret information in anti-retaliation lawsuit.--An individual who files a lawsuit for retaliation by an employer for reporting a suspected violation of law may disclose the trade secret to the attorney of the individual and use the trade secret information in the court proceeding, if the individual--

``(A) files any document containing the trade secret under seal; and

``(B) does not disclose the trade secret, except pursuant to court order.

``(3) Notice.--

``(A) In general.--An employer shall provide notice of the immunity set forth in this subsection in any contract or agreement with an employee that governs the use of a trade secret or other confidential information.

``(B) Policy document.--An employer shall be considered to be in compliance with the notice requirement in subparagraph (A) if the employer provides a cross-reference to a policy document provided to the employee that sets forth the employer's reporting policy for a suspected violation of law.

``(C) Non-compliance.--If an employer does not comply with the notice requirement in subparagraph (A), the employer may not be awarded exemplary damages or attorney fees under subparagraph (C) or (D) of section 1836(b)(3) in an action against an employee to whom notice was not provided.

``(D) Applicability.--This paragraph shall apply to contracts and agreements that are entered into or updated after the date of enactment of this subsection.

``(4) Employee defined.--For purposes of this subsection, the term `employee' includes any individual performing work as a contractor or consultant for an employer.

``(5) Rule of construction.--Except as expressly provided for under this subsection, nothing in this subsection shall be construed to authorize, or limit liability for, an act that is otherwise prohibited by law, such as the unlawful access of material by unauthorized means.''.

(b) Technical and Conforming Amendment.--Section 1838 of title 18, United States Code, is amended by striking ``This chapter'' and inserting ``Except as provided in section 1833(b), this chapter''.

Approved May 11, 2016.

Appendix B
Important Trade Secret Cases

Appendix B.1

The first trade secrets case in the United States also established trade secrets as assets.

John Vickery versus Jonas Welch.
SUPREME COURT OF MASSACHUSETTS, NORFOLK
36 Mass. 523; 1837 Mass. LEXIS 124;
19 Pick. 523
October, 1837, Decided

PRIOR HISTORY: [**1] This was an action of debt on a bond, dated August 11th, 1836. The case was tried before Morton J.

The condition of the bond was, that the defendant should, upon the request of the plaintiff on or before the 1st of September, 1836, convey and assure unto the plaintiff certain tenements and water privileges situate in Braintree, known by the name of Welch's chocolate-mills, "together with his exclusive right and art or secret manner of making chocolate, and all information pertaining to his said manner of making chocolate, with all movable furniture, apparatus and utensils, with the privilege of using the pans as they now are and continuing the same, by good and sufficient deed of warranty, freed and discharged from all incumbrances whatsoever," the plaintiff having first paid or secured to the defendant the purchase money, to wit, $ 2000 on the delivery of the deed and $ 7500 in eight annual payments as specified in the bond.

After the execution of the bond, three other persons agreed with the plaintiff to associate themselves with him in the purchase, and it was verbally agreed by the plaintiff and defendant, that the conveyance should be made to the plaintiff and his associates. [**2] On the 1st of September the plaintiff and his associates, and the defendant, convened at Boston for the purpose of completing the contract set forth in the bond. A tender of $ 2000, and of security for the balance of the purchase money, was duly made to the defendant, and a conveyance conformable to the condition was demanded of him. The defendant thereupon tendered a deed to the plaintiff and his associates, of the defendant's mills, machinery, tools and fixtures, which was objected to, and he was requested to make a conveyance of his right or art of making chocolate, in the words of the bond. This he refused to do, but said that he would communicate his secret manner of making chocolate and all he knew about it, but that he would not bind himself not to communicate it to others; stating that he had no exclusive patent or right, and that the plaintiff knew this at the time of the making of the bond. The defendant then returned to his home in Braintree. In the afternoon of the same day the plaintiff went to the defendant's house, and again tendered him $ 2000 and the security which he had tendered at Boston, having previously informed the defendant that the plaintiff's associates had [**3] declined to be further concerned with him in the purchase, and also demanded of the defendant a conveyance to the plaintiff alone, according to the condition of the bond. The plaintiff offered to have such a conveyance written, or to wait till the next morning for the defendant to procure one to be written. The defendant refused to give any other deed of the real estate than that which he had offered in Boston. That deed he again tendered to the plaintiff, and also a deed as follows: "Know all men &c. that I, Jonas Welch, of Braintree, in consideration of the contract executed between me and John Vickery, and in compliance therewith, do hereby convey and assure to said Vickery, his representatives and assigns, my exclusive right and arts or secret manner of manufacturing chocolate; hereby declaring, however, that I have no patent or other exclusive right or arts except what I have gained by my skill and experience, and never have professed or claimed to have any other; and I never have, and do not hereby even impliedly covenant not to communicate the results of my experience to others." He said that he could give no other instrument, and that he acted under advice of counsel. It appeared [**4] that two or three other persons had knowledge of the defendant's secret art of making chocolate, but that he had exacted an oath or a bond of them, not to divulge it while he should continue in the manufacture.

The cause was withdrawn from the jury, by consent of parties. If on these facts the whole Court should be of opinion, that a breach of the condition of the bond was proved, for which the defendant was liable, he was to be defaulted, and such a hearing to be had as to damages, as the Court should order.

DISPOSITION: Bond forfeited by the defendant, and he heard in chancery touching the damages.

HEADNOTES: Under a bond to convey and to assure to the obligee the obligor's chocolate-mill, "together with his exclusive right and art or secret manner of making chocolate and all information pertaining to his said manner of making chocolate," it is incumbent on the obligor to convey such exclusive right, with a covenant that he will communicate all the information necessary to enable the obligee to use the right or secret art, and that he will not divulge the secret to any other person.

Such a contract is not in restraint of trade.

COUNSEL: Choate, Crowninshield and Breck, for the plaintiff, contended, that the meaning of the contract was, that the plaintiff should have the exclusive benefit of making chocolate in the mode used by the defendant; that the defendant ought to have covenanted that he would communicate his secret art to the plaintiff and would not divulge it to any other person; and that at the least he should have tendered a deed in the very words of the bond. Hopkins v. Young, 11 Mass. 302; Propert v. Parker, 3 Mylne & Keen, 280; Browning v. Wright, 2 Bos. & Pul. 22. As to secrets of art, not patented, they referred [**5] to Phillips on Patents, 333, 340; Smith v. Dickenson, 3 Bos. & Pul. 630.

Richardson and Leland, for the defendant, said that a conveyance in the words of the bond, would not benefit the plaintiff; that the defendant agreed to impart to him a secret, and this he offered to do, but that he had not stipulated that he would not divulge the secret to others; and that such a stipulation would have been invalid, as being in restraint of trade. Com. Dig. Trade, D 3; 2 Wms's Saund. 156, note 1; Mitchel v. Reynolds, 1 P. Wms. 183; Davis v. Mason 5 T. R. 118; Prugnell v. Goss, Aleyn, 67; Alger v. Thacher, ante, p. 51.

Metcalf, in reply, cited to the point, that such secrets did not come within the principle of restraint of trade, Bryson v. Whitehead, 1 Sim. & Stuart, 74.

JUDGES: Putnam, J. delivered the opinion of the Court.

OPINIONBY: Putnam

OPINION:

[*525] Putnam J. delivered the opinion of the Court. The defendant contends that he has performed or offered to perform all that was required of him in the condition of the bond; while the plaintiff, on the other hand, denies any such performance or offer; and the question is to be settled by determining what was the true intent and meaning of [**6] the bond.

It is very clear that both parties had reference to the defendant's exclusive and secret art of making chocolate, and that art was to be conveyed to the plaintiff, together with the chocolate mills, &c. It must be taken that such transfer entered greatly into the value or consideration which the plaintiff stipulated to give for the property. The defendant was to transfer to the plaintiff, for his exclusive use, the secret manner which the defendant had of making chocolate. It must [*526] have been represented and understood by the parties, that the defendant had used such an exclusive art, which had given great advantage to him in the manufacture of chocolate. The plaintiff was to become the proprietor of the mills, and also of the secret mode of manufacture which the bond supposes was used and possessed by the defendant. The defendant was to sell, the plaintiff was to buy. Now we cannot perceive the least reason which, after such sale, would enable the defendant lawfully to retain any right in the property or rights sold, nor any right to convey to strangers, any part of what was to be transferred to the plaintiff. The exclusive right was to be transferred to the plaintiff, [**7] and we cannot conceive that it would be exclusive, if the defendant might, after such transfer, admit as many persons to participate as would pay for, or receive gratuitously, the same privilege which the defendant had granted or stipulated to grant.

Nor is there any difficulty in making a grant of this right, with proper covenants touching the same, in the deed conveying the real estate. It would have been sufficient, after describing the real estate, to have added, "together with the grantor's exclusive right and art or secret manner of making chocolate, and all information pertaining to his said manner of making chocolate," in the words of the condition of the bond, with a covenant (which the bond necessarily implies) in the nature of further assurance, to communicate all the information necessary to enable the plaintiff to have, use and enjoy the right or secret art exclusively, and a covenant that the plaintiff should from thenceforth have, use and enjoy the same right or secret art exclusively, without any lawful interference from any person or persons whomsoever.

It would not have been necessary to spread the secret at large in the conveyancing and place the same upon the [**8] public records of the county. That proceeding, as everybody sees, would have defeated the very transfer of the right. It was known to both parties, that there was no patent right granted to the defendant. It was for his exclusive secret that the parties treated, and the secret would not be kept if put at large upon the records.

But there was no difficulty in assuring this right to the plaintiff [*527] in general and apt words, with covenants to do all things, and give information in all things and matters, necessary to enable the plaintiff to use and enjoy the right; and that instruction and information should have been given to the plaintiff in private, to the end that he might preserve the right, that he might keep the secret, for his own use and exclusive enjoyment.

But the defendant has refused to do so. He has given such a construction as would enable him, for love or money, to communicate the secret to all other people; a construction which we think entirely inconsistent with his obligation to the plaintiff, and he has refused to assure the right to the plaintiff according to the true intent and meaning of the condition of the bond, although the plaintiff has done all [**9] things incumbent upon him to do as conditions precedent to such assurance.

It was contended for the defendant, that this obligation was void as being in restraint of trade. But we cannot suppose that the case comes within that doctrine. *Bryson* v. *Whitehead*, 1 Sim. & Stu. 74. The defendant claims to operate by a secret art. The public are not prejudiced by the transfer of it to the plaintiff. If it were worth any thing, the defendant would use the art and keep it secret, and it is of no consequence to the public whether the secret art be used by the plaintiff or by the defendant.

We are also of opinion, that there was a breach of the bond when the defendant refused, at his house in Braintree, to convey to the plaintiff himself. The verbal agreement which the plaintiff had made with his associates, had no effect to alter or change the liability of the defendant on his bond to the plaintiff, inasmuch as he, the defendant, had refused to assure the right to the plaintiff and his associates, and they, in consequence of such refusal, abandoned the enterprise.

It is the opinion of the whole Court, that the bond has been forfeited by the defendant, and that he may be heard in chancery [**10] touching the damages.

Appendix B.2

Failure to take reasonable measures to maintain the secrecy of information results in a forfeiture of trade secret rights in the information.

Omega Optical, Inc. v. Chroma Technology Corporation, Richard Stewart, et al.
No. 99-566
SUPREME COURT OF VERMONT
174 Vt. 10; 800 A.2d 1064; 2002 Vt. LEXIS 58
April 12, 2002, Filed
SUBSEQUENT HISTORY: [***1] As Corrected July 1, 2002. Motion for Reargument Denied May 10, 2002.

PRIOR HISTORY: On Appeal from Windham Superior Court. Robert Grussing III, J.

DISPOSITION: Affirmed.

COUNSEL: Bernard D. Lambek and Patricia K. Turley of Zalinger Cameron & Lambek, P.C., Montpelier, and R. Mark Halligan, Philip D. Segrest, Jr. and Steven E. Feldman of Welsh & Katz, Ltd., Chicago, Illinois, for Plaintiff-Appellant.

Craig Weatherly of Gravel & Shea, Burlington, Richard H. Munzing of Weber, Perra & Munzing, P.C., Brattleboro, and Heidi E. Harvey, Blair L. Perry, Jolynn M. Lussier and Michael E. Zelinger of Fish & Richardson P.C., Boston, Massachusetts, for Defendants-Appellees.

JUDGES: PRESENT: Dooley, Morse, Johnson and Skoglund, JJ., and Allen, C.J. (Ret.), Specially Assigned.

OPINIONBY: MORSE

OPINION: [**1065] [*12]

MORSE, J. Omega Optical appeals from a judgment of the superior court in favor of defendant Chroma Technology Corporation and several other named defendants in Omega's action for trade secret misappropriation, conversion, breach of loyalty, tortious interference with business relations, unfair competition, conspiracy and breach of contract. Omega argues on appeal: [***2] (1) the court erred in its determination of the proof necessary to demonstrate trade secret misappropriation, and it is entitled to judgment as a matter of law on that claim; (2) the court similarly erred on each of its remaining claims, with the exception of its claims for breach of contract and conversion, and it is entitled to judgment in its favor on each of the claims; (3) the court erroneously determined that Omega's proof of damages was not sufficiently definite so as to allow for an award without resorting to speculation or conjecture; (4) Omega is entitled to an award of punitive damages as a matter of law; and, finally, (5) the court failed to make adequate findings regarding its crediting of the individual defendants' testimony in light of Omega's theory and requested findings that a joint defense agreement undermined the credibility of that testimony. We affirm.

This case arises out of events spanning several months starting in early 1991, in which a number of Omega Optical employees left the company and went into business together under the name of Chroma Technology Corporation. Chroma began making thin-film optical interference filters used in fluorescence microscopy, [***3] a product that Omega had developed and also produces. On October 1, 1996, Omega [**1066] brought suit against Chroma and ten of its employees. Following a twenty-two-day bench trial, the trial court issued a 111-page decision finding in favor of defendants on all of Omega's claims.

Omega's appeal centers on its argument that, because defendants acquired substantial amounts of information that the court found was "protectible as a trade secret," it is entitled to judgment as a matter of law, notwithstanding the trial court's extensive findings that Omega failed to take reasonable steps to protect the information. The court [*13] concluded under the evidence that defendants owed no duty of confidentiality with regard to the information.

As a preliminary matter, and as Omega points out, because the operative facts of this case occurred before July 1, 1996, the effective date of the Vermont Trade Secrets Act, the common law predating the act applies to this case. See 9 V.S.A. § 4609 (providing the VTSA does not apply to misappropriation occurring before its effective date); see also McClary v. Hubbard, 97 Vt. 222, 229, 122 A. 469, 473 (1923) (recognizing right of [***4] action for misappropriation of a trade secret). But as the trial court noted, the act and the Restatement (Third) of Unfair Competition provide guidance with regard to Omega's common law claim for misappropriation of trade secrets.

In general, liability for trade secret misappropriation in the employment context requires proof of both the existence of a trade secret as well as unauthorized disclosure or use of the secret in breach of a duty of confidence. Aries Info. Sys., Inc. v. Pacific Mgmt. Sys. Corp., 366 N.W.2d 366, 368 (Minn. Ct. App. 1985); Restatement (Third) of Unfair Competition § § 40, 42 (1995). Employees, whether current or former, have a duty not to use or disclose confidential information imparted to them by their employer. Restatement (Second) of Agency § 396(b) (1958); see also Restatement (Third) of Unfair Competition § 42 cmt. b (former employee's duty of loyalty includes duty not to disclose employer's confidential information to others). The former employee's duty of confidence attaches to any information the employee knows or has reason to know is confidential. Restatement (Third) of Unfair Competition § 42 cmt. c; see also A.F.A. Tours, Inc. v. Whitchurch, 937 F.2d 82, 89 (2d Cir. 1991) [***5] (employer must take appropriate precautions to alert employee of need to maintain confidentiality of information subject to trade secret protection); Mercer v. C.A. Roberts Co., 570 F.2d 1232, 1238 (5th Cir. 1978) (agreement to keep employer's confidences implied where employee knows or should know based on particular circumstances that employer desired information to remain secret); Aries Info. Sys., 366 N.W.2d at 369 ("A duty of employer/employee confidentiality can arise at common law if the employee is given notice of what material is to be kept confidential."); Electro-Craft Corp. v. Controlled Motion, Inc., 332 N.W.2d 890, 903 (Minn. 1983) ("[A] common law duty of confidentiality arises out of the employer-employee relationship only as to information which the employer has treated as secret."). Whether an employee knows or should know certain information obtained from the employer is confidential can be implied from the totality of the [*14] circumstances; no explicit notice to the employee is necessarily required. In re Innovative Constr. Sys., Inc., 793 F.2d 875, 883 (7th Cir. 1986); Standard Brands, Inc. v. Zumpe, 264 F. Supp. 254, 262 (E.D. La. 1967); [***6] Sun Dial Corp. v. Rideout, 29 N.J. Super. 361, 102 A.2d 90, 96 (N.J. Super. Ct. App. Div. 1954); see also Restatement (Third) of Unfair Competition § 42 cmt. c ("If an employer establishes ownership of a trade secret [**1067] and circumstances sufficient to put the employee on notice that the information is confidential, the employment relationship will ordinarily justify the recognition of a duty of confidence.") (emphasis added). It is a fact-specific inquiry.

Omega argues, however, that employees who acquire valuable information in the course of their employment owe a duty of confidentiality to the employer merely by virtue of their status as employees, regardless of whether the employer has done anything either to protect the information or to communicate to the employees the confidential and proprietary nature of the information. This argument is simply at odds with the case law, which requires something more than the mere employer-employee relationship to establish a duty of confidentiality. See, e.g., Mercer, 570 F.2d at 1238 (noting "not all employment relationships are confidential," and determining that, although employee acquired intimate knowledge of [***7] his employer's operations, he owed no duty of confidentiality because he had not been informed that the information was secret and under the circumstances could have reasonably assumed it was not); Aries Info. Sys., 366 N.W.2d at 369; Electro-Craft Corp. v. Controlled Motion, Inc., 332 N.W.2d at 903. As noted above, whether a duty of confidence attaches is a factual inquiry. Furthermore, our decision in McClary v. Hubbard makes plain that the party claiming a trade secret must demonstrate that it has taken steps to ensure the information's secrecy to prevail on a claim for trade secret misappropriation. McClary, 97 Vt. at 232-33, 245, 122 A. at 473, 479.

In analyzing the facts of this case, the trial court applied the Restatement's approach to Omega's claim for trade secret misappropriation. In other words, it determined that the body of knowledge necessary to produce the thin-film optical interference filters was "sufficiently valuable, and not generally well known, that it [was] protectible as a trade secret." n1 Consistent with McClary, the [*15] court went on, however, to conclude that Omega had failed to take measures toward protecting the information, [***8] and the circumstances under which the employees acquired the information failed to indicate to them that the information was confidential. Therefore, the defendants owed no duty of confidentiality to Omega, and their use of the valuable information in their new venture did not constitute misappropriation of that information.

n1 Although similar, the VTSA and the Restatement take somewhat different approaches to analyzing a claim for misappropriation of trade secrets. As we noted in Dicks v. Jensen, 172 Vt. 43, 768 A.2d 1279, 1282 (2001), the VTSA includes in its definition of "trade secret" a requirement that the information be the subject of reasonable efforts to maintain its secrecy. 9 V.S.A. § 4601(3)(B). If such efforts have not been made, the information is not a "trade secret." On the other hand, the Restatement's definition does not incorporate such a requirement. Restatement (Third) of Unfair Competition § 39. Rather, its definition merely parallels the first part of our statutory definition. Compare id., with 9 V.S.A. § 4601(3)(A). Under the Restatement, information is a "trade secret" if it is used and valued in the operation of a business and is sufficiently secret, i.e., not generally well known, that it affords economic advantage to the business. Restatement (Third) of Unfair Competition § 39. Notably, the "secret" element of the Restatement definition addresses itself only to the question of whether the information bestows any competitive advantage on the employer.

[***9]

Specifically with respect to the circumstances under which the defendants acquired the information, the trial court found that while defendants were employed at Omega, the company had "no internal policies concerning confidentiality, nondisclosure or noncompetition." The court found that an open atmosphere prevailed [**1068] in which employees were encouraged to share information about the development and production of filters. In contrast, during this same time, the company did have confidentiality policies with regard to proprietary information belonging to Omega's customers. The court found that Omega failed to convey during the defendants' course of employment the confidential nature of the information it now seeks to protect and, when Omega did institute a policy of confidentiality, the defendants still employed with the company refused to sign onto it and resigned. The court also found that Omega had few security measures in place prior to defendants leaving the company and the few security measures that did exist, such as signing out keys, were not enforced or even monitored. For a significant period of time, members of the public had access to work areas. The court also noted that [***10] many of the measures cited by Omega as security measures in the course of the trial actually evolved for different purposes and Omega's claims, such as its location as a security measure, were post hoc rationalizations.

These findings are supported by the record. For instance, Robert Johnson, Omega's founder, testified the company had no lock on the door of its first facility, and at the next facility, the doors connecting [*16] Omega to other businesses were not locked. He testified that Omega avoided traditional security measures, and that the company was relying on the fact that its employees were "intelligent people" who would intuit that the information they worked with should be kept confidential. He also testified that he expected employees to know that information acquired in the course of their employment was proprietary and confidential simply based on the nature of the information. He indicated that Omega did not mark any of its own documents "confidential" because it would have undercut the fact that all documents at Omega were confidential. Cf. J.T. Healy & Son, Inc. v. James A. Murphy & Son, Inc., 357 Mass. 728, 260 N.E.2d 723, 730 (Mass. 1970) (concluding that [***11] employer's claimed security measure of doing nothing in order not to "excite undue interest" in information was likely only an afterthought and was "at variance with the rule that individuals must be constantly admonished that a process is secret and must be kept so").

There was also testimony and other documentary evidence that (1) Omega had no written policy of confidentiality; (2) just prior to the time defendants were contemplating leaving Omega, the company recognized the need to establish guidelines to help employees distinguish between confidential and nonconfidential information; (3) very little awareness existed among employees as to what the company considered proprietary; and (4) record keeping at Omega was sloppy, in some cases nonexistent, and the company's technical information was not kept in an organized or centralized, controlled manner. In one instance, defendant Kebbel testified that after Omega donated one of its computers to a local day-care center, Dr. Johnson asked Kebbel if the company still had the computer because it contained information that Dr. Johnson needed. In light of this and other evidence available to the trial court, we find no error in the court's [***12] finding that Omega failed to take steps to put its employees on explicit or implicit notice that certain information conveyed to them during their employment was to be kept confidential. Hence, defendants did not receive the information in confidence and, therefore, did not breach their duty of confidence to Omega, their former employer.

Omega argues, however, that the court erred by examining the circumstances surrounding the acquisition of the [**1069] information when determining whether the defendants owed a duty of confidentiality. Omega contends that it is the time of use that is determinative [*17] regarding whether former employees owe a duty of confidentiality to a former employer. Omega confuses the relevant times for determining whether a duty exists and whether the duty has been breached, that is, whether the information has been misused or misappropriated. To determine whether a duty of confidentiality arises, courts examine whether the information was acquired under circumstances that would indicate to an individual that the information is confidential, while courts look to the time of use to determine whether the individual knows that the information being used is subject to a duty [***13] of confidentiality such that the use constitutes a breach of that duty and, thus, misappropriation. Compare Restatement (Third) of Unfair Competition § 41(b)(1) (noting person owes other party a duty of confidentiality with respect to a trade secret if the trade secret was disclosed to that person under circumstances that would justify the conclusion that, "at the time of the disclosure, the person knew or had reason to know that the disclosure was intended to be in confidence") (emphasis added), with id. at § 40(b)(1) (stating that individual is liable for appropriation of a trade secret if, among other things, "the actor uses or discloses the other's trade secret without the other's consent and, at the time of the use or disclosure, the actor knows or has reason to know that the information is a trade secret that the actor acquired under circumstances creating a duty of confidence owed by the actor to the other") (emphasis added); see also Electro-Craft Corp., 332 N.W.2d at 901 ("Trade secret protection . . . depends upon a continuing course of conduct by the employer, a course of conduct which creates a confidential relationship.") (emphasis added).

The trial court's [***14] findings are supported by the record, and the trial court correctly applied the governing law. Consequently, we discern no error in the trial court's judgment on the claim for trade secret misappropriation that would require either reversal or entry of judgment in Omega's favor. See Highgate Assocs. v. Merryfield, 157 Vt. 313, 315-16, 597 A.2d 1280, 1281-82 (1991) ("Where the trial court has applied the proper legal standard, we will uphold its conclusions of law if reasonably supported by its findings.").

On its related claims of breach of loyalty, tortious interference with business relations, unfair competition and conspiracy, Omega asserts that the trial court made errors of law that not only require de novo review, but that, once remedied, entitle Omega to judgment in its favor on each of the claims. As we noted in Dicks v. Jensen, the Vermont Trade Secrets Act explicitly supplants common law tort claims that provide civil remedies for the misappropriation of trade secrets. 172 Vt. at 51, [*18] 768 A.2d at 1285; see also 9 V.S.A. § 4607. Therefore, if the VTSA governed this case, arguably at least some of Omega's claims would be [***15] eliminated. As noted above, however, the acts in question in this case took place before the effective date of the VTSA, and therefore the common law still governs.

With regard to its assertion that the trial court's decision on each of these claims should be reviewed de novo, Omega fails to point to any individual error of law by the trial court on any of the claims and instead reargues the evidence before the trial court on each claim, cataloguing the evidence in its favor. The standard of review in such circumstances is a determination whether the court's findings of fact are supported by the record and whether those findings reasonably support its conclusions. [**1070] Highgate, 157 Vt. at 315-16, 597 A.2d at 1281-82. In making that determination, we disregard any mitigating evidence and view the record in the light most favorable to the trial court's findings. Id. at 315, 597 A.2d at 1281 ("A finding will not be disturbed merely because it is contradicted by substantial evidence; rather, an appellant must show there is no credible evidence to support the finding.").

To the degree that the additional claims are premised on Omega's claim that defendants misappropriated [***16] Omega's trade secrets, the trial court's disposition of that claim also disposes of the additional claims.

Furthermore, as the trial court noted with regard to Omega's breach of loyalty claim, courts have generally held that at-will employees may plan to compete with their employer even while still employed there and may freely compete with the employer once they are no longer employed there. See Augat, Inc. v. Aegis, Inc., 409 Mass. 165, 565 N.E.2d 415, 419 (Mass. 1991); Metal Lubricants Co. v. Engineered Lubricants Co., 411 F.2d 426, 429-30 (8th Cir. 1969) (noting that employees may plan to compete with their employer while still in its employ, and "absent [a] covenant not to compete or breach of a confidential relationship, an employee is free to leave his employment and enter into competition with his former employer"). This behavior does not constitute a breach of a duty of loyalty. Such at-will employees are restricted, however, from misappropriating trade secrets and soliciting customers for their new venture while still employed at the former employer. See Augat, 565 N.E.2d at 419 (noting that an at-will employee may plan to go into [***17] competition with his or her employer while still employed, but that the employee may not solicit the employer's customers while still employed there and may not appropriate the employer's trade secrets or other confidential [*19] information). The trial court specifically found that, while defendants did formulate plans for the creation of Chroma while still working at Omega, they did not solicit Omega's customers while still employed there and continued to perform their duties at Omega in good faith. As these findings are supported by the record, we will not disturb them on appeal.

Omega argues, however, that competition for Omega customers by defendants following their departure from the company constitutes a breach of the duty of loyalty as a matter of law. But as we noted with respect to customer lists in Dicks, when an employer does not take steps to protect information such as customer lists, competition for those customers by a former employee after that employee has left the company is legitimate. Dicks, 172 Vt. at 51, 768 A.2d at 1285.

Omega also makes a more general argument that "as former employees of Omega, the individual defendants each . . . continued to owe [***18] Omega a duty of loyalty, which included the duty to refrain from acting for their own benefit or the benefit of Chroma to the detriment of Omega." (Emphasis added.) Omega cites no authority for the proposition that at-will employees continue to owe a duty of loyalty to a former employer, even after they have left that employment, that constrains them from ever acting to the detriment of that employer. Such a common law duty would prevent an employee from ever going to work for a competitor even in the absence of an agreement not to do so, an anomalous result. Cf. Dicks, 172 Vt. at 51, 768 A.2d at 1285 (given restraint with which this Court enforces explicit noncompete agreements, we declined to imply one).

On Omega's claim for tortious interference with business relations, the trial court concluded that, although defendants may have used aggressive sales techniques, [**1071] Omega failed to demonstrate that the defendants acted with the sole purpose of harming Omega, used improper means in competing with Omega for customers, or used criminal or fraudulent means to advance their own competitive interests. See Gifford v. Sun Data, Inc., 165 Vt. 611, 613, 686 A.2d 472, 474-75 (1996) [***19] (mem.) (noting proof necessary to prevail on claim for tortious interference with prospective contractual relations and stating "competitive business practices are not tortious"). The trial court also concluded that Omega had failed to establish a causal connection between any of defendants' practices and a loss of business at Omega. Although Omega cites to evidence it argues demonstrates that defendants used improper means in competing for customers, its brief [*20] fails to address the trial court's conclusion that it failed in its proof on the causal element of the tort. Accordingly, we will not reverse the trial court on this claim.

Omega also claims that the trial court ignored its evidence of customer confusion with regard to its claim of unfair competition, citing a single incident in which a representative of a company that distributes microscopes indicated that he thought Omega was a trade name for Chroma's product, leading him to contact Chroma when he meant to contact Omega about ordering a filter set. Omega points out the trial court failed to address its proposed findings regarding this incident and its proposed finding regarding Chroma's failure to dispel the representative's [***20] confusion about the companies. The trial court, however, explicitly acknowledged that "some confusion [of the companies] was inevitable" given that both manufactured the same product and were located in Brattleboro. It went on to conclude, though, that defendants had not acted with a conscious object of fostering confusion, the names of the two companies were not so similar as to inevitably lead to customer confusion and Omega had failed to establish the existence of "any significant level of confusion" among the companies' customers. Omega's citation to evidence of the isolated incident above does not undermine these conclusions.

Omega's argument that it is entitled to judgment on its conspiracy claim rests on its argument that the defendants acted in concert in engaging in the behavior giving rise to its other claims, including that for trade secret misappropriation. As these other claims are without merit, so is its claim for conspiracy. In sum, we decline Omega's invitation to reverse the trial court's judgment on each of the claims for breach of loyalty, tortious interference with business relations, unfair competition and conspiracy, and enter judgment in its favor.

Because [***21] of our disposition, we need not address Omega's arguments on the damages issues. Finally, with respect to Omega's argument that the trial court failed to make adequate findings on the credibility of each of the individual defendants and erred by failing to adopt instead Omega's proposed findings on the issue of their credibility, we discern no reversible error. Determinations of credibility are solely the province of the factfinder, Cabot v. Cabot, 166 Vt. 485, 497, 697 A.2d 644, 652 (1997) (noting it is the province of the trial court to make determinations of witness credibility); Mullin v. Phelps, 162 Vt. 250, 261, 647 A.2d 714, 720 (1994) (noting that in reviewing findings by a [*21] trial court, this Court is not to "reweigh evidence or . . . make findings of credibility de novo"), and the failure to adopt a party's proposed findings on the issue of witness credibility is not grounds for reversal. See McCormick v. McCormick, 150 Vt. 431, 435, 553 A.2d 1098, 1101 (1988) (noting that failure [**1072] of court to adopt party's proposed findings is not cause for reversal because court is free to choose the evidence it finds persuasive). Furthermore, the trial [***22] court is not obligated to explain to a party why it is not adopting the party's proposed findings. Thus, the trial court's failure to adopt Omega's theory that defendants' joint defense agreement rendered their testimony not worthy of belief or to explain why the joint defense agreement did not undermine their credibility does not constitute grounds for reversal.

Affirmed.

Appendix B.3

Judge Richard Posner emphasized the importance of trade secrets to America's economy, while setting limits on the extent of secrecy measures required to be considered reasonable under the circumstances.

ROCKWELL GRAPHIC SYSTEMS, INCORPORATED, Plaintiff-Appellant, v. DEV INDUSTRIES, INCORPORATED; PRESS MACHINERY CORPORATION; and ROBERT FLECK, Defendants-Appellees
No. 90-1499
UNITED STATES COURT OF APPEALS FOR THE SEVENTH CIRCUIT
925 F.2d 174; 1991 U.S. App. LEXIS 1969; 17 U.S.P.Q.2D (BNA) 1780
January 11, 1991, Argued
February 11, 1991, Decided

PRIOR HISTORY: [**1] Appeal from the United States District Court for the Northern District of Illinois, Eastern Division. No. 84 C 6746; Ann Claire Williams, Judge.

DISPOSITION:
Reversed And Remanded.

COUNSEL:
Michael O. Warnecke, William P. Oberhardt, Deborah S. Ruff, John M. Augustyn, Neuman, Williams, Anderson & Olson, Chicago, Illinois, Richard A. Speer, Pittsburgh, Pennsylvania, R. Paul Eck, Cicero, Illinois, for Plaintiff-Appellant.

Louis B. Garippo, Lydon & Griffin, Chicago, Illinois, Stephen P. Carponelli, James E. Hussey, Carponelli, Krug & Adamski, Chicago, Illinois, James J. Flynn, Quinn, Jacobs, Barry & Miller, Chicago,

Illinois, for Defendant-Appellee.

Lawrence S. Wick, Leydig, Voit & Mayer, Chicago, Illinois, Frederick T. Stocker, Manufacturers' Alliance for Productivity and Innovation, Inc, Washington, District of Columbia, for Amicus Curiae.

JUDGES:

Cummings, Posner, and Flaum, Circuit Judges.

OPINIONBY:

POSNER

OPINION:

[*175] POSNER, Circuit Judge

This is a suit for misappropriation of trade secrets. Rockwell Graphic Systems, a manufacturer of printing presses used by newspapers, and of parts for those presses, brought the suit against DEV Industries, a competing manufacturer, and against the president of DEV, who used to be employed by Rockwell. The case is in federal [**2] court by virtue of the RICO ("Racketeer Influenced and Corrupt Organizations") statute. 18 U.S.C. § § 1961 *et seq.* The predicate acts required for liability under RICO are acts of misappropriation (and related misconduct, such as alleged breaches of fiduciary duty) committed by the individual defendant, Fleck, and by another former employee of Rockwell and present employee of DEV, Peloso. These acts are alleged to violate Illinois law, and in pendent counts Rockwell seeks to impose liability for them directly under that law as well as indirectly under RICO. The district judge granted summary judgment for the defendants upon the recommendation of a magistrate who concluded that Rockwell had no trade secrets because it had failed to take reasonable precautions to maintain secrecy. (730 F. Supp. 171). Therefore there had been no misappropriation, which in turn was the foundation for the predicate acts; so the RICO count had to be dismissed. With the federal claim out of the case, the district judge relinquished jurisdiction over the pendent counts, resulting in a dismissal of the entire case. 730 F. Supp. 171 (1990).

When we said that Rockwell manufactures both printing presses and replacement parts for its presses — "wear parts" [**3] or "piece parts," they are called — we were speaking approximately. Rockwell does not always manufacture the parts itself. Sometimes when an owner of one of Rockwell's presses needs a particular part, or when Rockwell anticipates demand for the part, it will subcontract the manufacture of it to an independent machine shop, called a "vendor" by the parties. When it does this it must give the vendor a "piece part drawing" indicating materials, dimensions, tolerances, and methods of manufacture. Without that information the vendor could not manufacture the part. Rockwell has not tried to patent the piece parts. It believes that the purchaser cannot, either by inspection or by "reverse engineering" (taking something apart in an effort to figure out how it was made), discover how to manufacture the part; to do that you need the piece part drawing, which contains much information concerning methods of manufacture, alloys, tolerances, etc. that cannot be gleaned from the part itself. So Rockwell tries — whether hard enough is the central issue in the case — to keep the piece part drawings secret, though not of course from the vendors; they could not manufacture the parts for Rockwell [**4] without the drawings. DEV points out that some of the parts are for presses that Rockwell no longer manufactures. But as long as the [*176] presses are in service — which can be a very long time — there is a demand for replacement parts.

Rockwell employed Fleck and Peloso in responsible positions that gave them access to piece part drawings. Fleck left Rockwell in 1975 and three years later joined DEV as its president. Peloso joined DEV the following year after being fired by Rockwell when a security guard caught him removing piece part drawings from Rockwell's plant. This suit was brought in 1984, and pretrial discovery by Rockwell turned up 600 piece part drawings in DEV's possession, of which 100 were Rockwell's. DEV claimed to have obtained them lawfully, either from customers of Rockwell or from Rockwell vendors, contrary to Rockwell's claim that either Fleck and Peloso stole them when they were employed by it or DEV obtained them in some other unlawful manner, perhaps from a vendor who violated his confidentiality agreement with Rockwell. Thus far in the litigation DEV has not been able to show which customers or vendors lawfully supplied it with Rockwell's piece part drawings.

The [**5] defendants persuaded the magistrate and the district judge that the piece part drawings weren't really trade secrets at all, because Rockwell made only perfunctory efforts to keep them secret. Not only were there thousands of drawings in the hands of the vendors; there were thousands more in the hands of owners of Rockwell presses, the customers for piece parts. The drawings held by customers, however, are not relevant. They are not piece part drawings, but assembly drawings. (One piece part drawing in the record is labeled "assembly," but as it contains dimensions, tolerances, and other specifications it is really a piece part drawing, despite the label.) An assembly drawing shows how the parts of a printing press fit together for installation and also how to integrate the press with the printer's other equipment. Whenever Rockwell sells a printing press it gives the buyer assembly drawings as well. These are the equivalent of instructions for assembling a piece of furniture. Rockwell does not claim that they contain trade secrets. It admits having supplied a few piece part drawings to customers, but they were piece part drawings of obsolete parts that Rockwell has no interest in [**6] manufacturing and of a safety device that was not part of the press as originally delivered but that its customers were clamoring for; more to the point, none of these drawings is among those that Rockwell claims DEV misappropriated.

The distinction between assembly and piece part drawings is not esoteric. *A.H. Emery Co. v. Marcan Products Corp.*, 268 F. Supp. 289, 300 (S.D.N.Y. 1967), aff'd, 389 F.2d 11, 16 (2d Cir. 1968), marks it, and along with other cases declares — what is anyway obvious — that a firm's act in making public some of its documents (or part of a document) does not destroy the status as trade secrets of information contained in other documents (or another part of the same document). *Alexander & Alexander, Inc. v. Drayton*, 378 F. Supp. 824, 833 (E.D. Pa. 1974), aff'd without opinion, 505 F.2d 729 (3d Cir. 1974); *Ecolaire Inc. v. Crissman*, 542 F. Supp. 196, 206 (E.D. Pa. 1982); *Laser Industries, Ltd. v. Eder Instrument Co.*, 573 F. Supp. 987, 991 (N.D. Ill. 1983). It is immaterial that Rockwell affixed the same legend enjoining the user to confidentiality to its assembly drawings as it did to its piece part drawings. Perhaps thinking of the doctrine [**7] of patent misuse (on which see *USM Corp. v. SPS Technologies, Inc.*, 694 F.2d 505, 510-12 (7th Cir. 1982), and cases cited there), DEV suggests that if a firm claims trade secret protection for information that is not really secret, the firm forfeits trade secret protection of information that is secret. There is no such doctrine — even the patent misuse doctrine does not decree forfeiture of the patent as the sanction for misuse — and it would make no sense. This is not only because there are any number of innocent explanations for Rockwell's action in "overclaiming" trade secret protection (if that is what it was doing) — such as an excess of caution, uncertainty as to the scope of trade secret protection, concern that clerical personnel will not always be able to distinguish between assembly and piece part [*177] drawings at a glance, and the sheer economy of a uniform policy — but also because it would place the owner of trade secrets on the razor's edge. If he stamped "confidential" on every document in sight, he would run afoul of what we are calling (without endorsing) the misuse doctrine. But if he did not stamp confidential on every document he would lay himself open to an [**8] accusation that he was sloppy about maintaining secrecy — and in fact DEV's main argument is that Rockwell *was* impermissibly sloppy in its efforts to keep the piece part drawings secret.

On this, the critical, issue, the record shows the following. (Because summary judgment was granted to DEV, we must construe the facts as favorably to Rockwell as is reasonable to do.) Rockwell keeps all its engineering drawings, including both piece part and assembly drawings, in a vault. Access not only to the vault, but also to the building in which it is located, is limited to authorized employees who display identification. These are mainly engineers, of whom Rockwell employs 200. They are required to sign agreements not to disseminate the drawings, or disclose their contents, other than as authorized by the company. An authorized employee who needs a drawing must sign it out from the vault and return it when he has finished with it. But he is permitted to make copies, which he is to destroy when he no longer needs them in his work. The only outsiders allowed to see piece part drawings are the vendors (who are given copies, not originals). They too are required to sign confidentiality agreements, [**9] and in addition each drawing is stamped with a legend stating that it contains proprietary material. Vendors, like Rockwell's own engineers, are allowed to make copies for internal working purposes, and although the confidentiality agreement that they sign requires the vendor to return the drawing when the order has been filled, Rockwell does not enforce this requirement. The rationale for not enforcing it is that the vendor will need the drawing if Rockwell reorders the part. Rockwell even permits unsuccessful bidders for a piece part contract to keep the drawings, on the theory that the high bidder this round may be the low bidder the next. But it does consider the ethical standards of a machine shop before making it a vendor, and so far as appears no shop has ever abused the confidence reposed in it.

The mere fact that Rockwell gave piece part drawings to vendors — that is, disclosed its trade secrets to "a limited number of outsiders for a particular purpose" — did not forfeit trade secret protection. *A.H. Emery Co. v. Marcan Products Corp.*, 389 F.2d 11, 16 (2d Cir. 1968). On the contrary, such disclosure, which is often necessary to the efficient exploitation of a trade [**10] secret, imposes a duty of confidentiality on the part of the person to whom the disclosure is made. *Jones v. Ulrich*, 342 Ill. App. 16, 25-26, 95 N.E.2d 113, 117 (1950); *Crocan Corp. v. Sheller-Globe Corp.*, 385 F. Supp. 251, 253 (N.D. Ill. 1974). But with 200 engineers checking out piece part drawings and making copies of them to work from, and numerous vendors receiving copies of piece part drawings and copying them, tens of thousands of copies of these drawings are floating around outside Rockwell's vault, and many of these outside the company altogether. Although the magistrate and the district judge based their conclusion that Rockwell had not made adequate efforts to maintain secrecy in part at least on the irrelevant fact that it took no effort at all to keep its assembly drawings secret, DEV in defending the judgment that it obtained in the district court argues that Rockwell failed to take adequate measures to keep even the piece part drawings secret. Not only did Rockwell not limit copying of those drawings or insist that copies be returned; it did not segregate the piece part drawings from the assembly drawings and institute more secure procedures for the former. [**11] So Rockwell could have done more to maintain the confidentiality of its piece part drawings than it did, and we must decide whether its failure to do more was so plain a breach of the obligation of a trade secret owner to make reasonable efforts to maintain secrecy as to justify the [*178] entry of summary judgment for the defendants.

The requirement of reasonable efforts has both evidentiary and remedial significance, and this regardless of which of the two different conceptions of trade secret protection prevails. (Both conceptions have footholds in Illinois law, as we shall see.) The first and more common merely gives a remedy to a firm deprived of a competitively valuable secret as the result of an independent legal wrong, which might be conversion or other trespass or the breach of an employment contract or of a confidentiality agreement. Under this approach, because the secret must be taken by improper means for the taking to give rise to liability, Ill. Rev. Stat. ch. 140, paras. 352(a), (b)(1), (2)(A), (B); Restatement of Torts § 757 (1939); *ILG Industries, Inc. v. Scott*, 49 Ill. 2d 88, 93, 273 N.E.2d 393, 396 (1971); *Brunswick Corp. v. Outboard Marine Corp.*, 79 Ill. 2d 475, 479, 404 N.E.2d 205, 207, 38 Ill. Dec. 781 [**12] (1980), the only significance of trade secrecy is that it allows the victim of wrongful appropriation to obtain damages based on the competitive value of the information taken. The second conception of trade secrecy, illustrated by *E.I. duPont de Nemours & Co. v. Christopher*, 431 F.2d 1012 (5th Cir. 1970), and in Illinois by Ill. Rev. Stat. ch. 140, para. 352(b)(2)(C), and *Schulenburg v. Signatrol, Inc.*, 33 Ill. 2d 379, 387-88, 212 N.E.2d 865, 869 (1965), is that "trade secret" picks out a class of socially valuable information that the law should protect even against nontrespassory or other lawful conduct — in *Christopher*, photographing a competitor's roofless plant from the air while not flying directly overhead and hence not trespassing or committing any other wrong independent of the appropriation of the trade secret itself. See also *Brunswick Corp. v. Outboard Marine Corp., supra*, 79 Ill. 2d at 479, 404 N.E.2d at 207; Restatement, *supra*, § 758(b).

Since, however, the opinion in *Christopher* describes the means used by the defendant as "improper," 431 F.2d at 1015-17, which is also the key to liability under the first, more conventional [**13] conception of trade secret protection, it is unclear how distinct the two conceptions really are. It is not as if *Christopher* proscribes *all* efforts to unmask a trade secret. It specifically mentions reverse engineering as a proper means of doing so. *Id.* at 1015. This difference in treatment is not explained, but it may rest on the twofold idea that reverse engineering involves the use of technical skills that we want to encourage, and that anyone should have the right to take apart and to study a product that he has bought.

It should be apparent that the two different conceptions of trade secret protection are better described as different emphases. The first emphasizes the desirability of deterring efforts that have as their sole purpose and effect the redistribution of wealth from one firm to another. The second emphasizes the desirability of encouraging inventive activity by protecting its fruits from efforts at appropriation that are, indeed, sterile wealth-redistributive — not productive — activities. The approaches differ, if at all, only in that the second does not limit the class of improper means to those that fit a preexisting pigeonhole in the law of tort or contract [**14] or fiduciary duty — and it is by no means clear that the first approach assumes a closed class of wrongful acts, either.

Under the first approach, at least if narrowly interpreted so that it does not merge with the second, the plaintiff must prove that the defendant obtained the plaintiff's trade secret by a wrongful act, illustrated here by the alleged acts of Fleck and Peloso in removing piece part drawings from Rockwell's premises without authorization, in violation of their employment contracts and confidentiality agreements, and using them in competition with Rockwell. Rockwell is unable to prove directly that the 100 piece part drawings it got from DEV in discovery were stolen by Fleck and Peloso or obtained by other improper means. But if it can show that the probability that DEV could have obtained them otherwise — that is, without engaging in wrongdoing — is slight, then it will have taken a giant step toward proving what it must prove in order to recover under the first theory of [*179] trade secret protection. The greater the precautions that Rockwell took to maintain the secrecy of the piece part drawings, the lower the probability that DEV obtained them properly and the higher [**15] the probability that it obtained them through a wrongful act; the owner had taken pains to prevent them from being obtained otherwise.

Under the second theory of trade secret protection, the owner's precautions still have evidentiary significance, but now primarily as evidence that the secret has real value. For the precise means by which the defendant acquired it is less important under the second theory, though not completely unimportant; remember that even the second theory allows the unmasking of a trade secret by *some* means, such as reverse engineering. If Rockwell expended only paltry resources on preventing its piece part drawings from falling into the hands of competitors such as DEV, why should the law, whose machinery is far from costless, bother to provide Rockwell with a remedy? The information contained in the drawings cannot have been worth much if Rockwell did not think it worthwhile to make serious efforts to keep the information secret.

The remedial significance of such efforts lies in the fact that if the plaintiff has allowed his trade secret to fall into the public domain, he would enjoy a windfall if permitted to recover damages merely because the defendant took [**16] the secret from him, rather than from the public domain as it could have done with impunity. *Brunswick Corp. v. Outboard Marine Corp., supra*, 79 Ill. 2d at 479, 404 N.E.2d at 207; *Van Products Co. v. General Welding & Fabricating Co.*, 419 Pa. 248, 267-68, 213 A.2d 769, 779-80 (1965) (repudiating the interpretation of Pennsylvania law that this court had adopted in *Smith v. Dravo Corp.*, 203 F.2d 369, 374-75 (7th Cir. 1953)). It would be like punishing a person for stealing property that he believes is owned by another but that actually is abandoned property. If it were true, as apparently it is not, that Rockwell had given the piece part drawings at issue to customers, and it had done so without requiring the customers to hold them in confidence, DEV could have obtained the drawings from the customers without committing any wrong. The harm to Rockwell would have been the same as if DEV had stolen the drawings from it, but it would have had no remedy, having parted with its rights to the trade secret. This is true whether the trade secret is regarded as property protected only against wrongdoers or (the logical extreme of the second conception, although no case — not even [**17] *Christopher* — has yet embraced it and the patent statute may preempt it) as property protected against the world. In the first case, a defendant is perfectly entitled to obtain the property by lawful conduct if he can, and he can if the property is in the hands of persons who themselves committed no wrong to get it. In the second case the defendant is perfectly entitled to obtain the property if the plaintiff has abandoned it by giving it away without restrictions.

It is easy to understand therefore why the law of trade secrets requires a plaintiff to show that he took reasonable precautions to keep the secret a secret. If analogies are needed, one that springs to mind is the duty of the holder of a trademark to take reasonable efforts to police infringements of his mark, failing which the mark is likely to be deemed abandoned, or to become generic or descriptive (and in either event be unprotectable). 1 McCarthy, Trademarks and Unfair Competition § 17:50, at pp. 778-80 (2d ed. 1984). The trademark owner who fails to police his mark both shows that he doesn't really value it very much and creates a situation in which an infringer may have been unaware that he was using a proprietary [**18] mark because the mark had drifted into the public domain, much as DEV contends Rockwell's piece part drawings have done.

But only in an extreme case can what is a "reasonable" precaution be determined on a motion for summary judgment, because the answer depends on a balancing of costs and benefits that will vary from case to case and so require estimation and measurement by persons knowledgeable in the particular field of endeavor involved. On the one hand, the more the owner of the [*180] trade secret spends on preventing the secret from leaking out, the more he demonstrates that the secret has real value deserving of legal protection, that he really was hurt as a result of the misappropriation of it, and that there really *was* misappropriation. On the other hand, the more he spends, the higher his costs. The costs can be indirect as well as direct. The more Rockwell restricts access to its drawings, either by its engineers or by the vendors, the harder it will be for either group to do the work expected of it. Suppose Rockwell forbids *any* copying of its drawings. Then a team of engineers would have to share a single drawing, perhaps by passing it around or by working in the same room, huddled [**19] over the drawing. And how would a vendor be able to make a piece part — would Rockwell have to bring all that work in house? Such reconfigurations of patterns of work and production are far from costless; and therefore perfect security is not optimum security.

There are contested factual issues here, bearing in mind that what is reasonable is itself a fact for purposes of Rule 56 of the civil rules. *Cooter & Gell v. Hartmarx Corp.*, 496 U.S. 384, 110 S. Ct. 2447, 2459, 110 L. Ed. 2d 359 (1990); *Mucha v. King*, 792 F.2d 602, 605 (7th Cir. 1986); *Nunez v. Superior Oil Co.*, 572 F.2d 1119, 1126 (5th Cir. 1978). Obviously Rockwell took some precautions, both physical (the vault security, the security guards — one of whom apprehended Peloso *in flagrante delicto*) and contractual, to maintain the confidentiality of its piece part drawings. Obviously it could have taken more precautions. But at a cost, and the question is whether the additional benefit in security would have exceeded that cost. We do not suggest that the question can be answered with the same precision with which it can be posed, but neither can we say that no reasonable jury could find that Rockwell had done enough and could then go on to [**20] infer misappropriation from a combination of the precautions Rockwell took and DEV's inability to establish the existence of a lawful source of the Rockwell piece part drawings in its possession.

This is an important case because trade secret protection is an important part of intellectual property, a form of property that is of growing importance to the competitiveness of American industry. Patent protection is at once costly and temporary, and therefore cannot be regarded as a perfect substitute. If trade secrets are protected only if their owners take extravagant, productivity-impairing measures to maintain their secrecy, the incentive to invest resources in discovering more efficient methods of production will be reduced, and with it the amount of invention. And given the importance of the case we must record our concern at the brevity of the district court's opinion granting summary judgment (one and a half printed pages). Brevity is the soul of wit, and all that, and the district judge did have the benefit of a magistrate's opinion; but it is vital that commercial litigation not appear to be treated as a stepchild in the federal courts. The future of the nation depends in no [**21] small part on the efficiency of industry, and the efficiency of industry depends in no small part on the protection of intellectual property.

The judgment is reversed and the case remanded to the district court for further proceedings consistent with this opinion (including reinstatement of the pendent counts).

REVERSED AND REMANDED.

Appendix B.4

The United States Court of Appeals for the Seventh Circuit had to consider the Illinois variation of the Uniform Trade Secret Act as well as the six factors from the Restatement (First) of Torts in reversing the district court's finding in this case.

LEARNING CURVE TOYS, INCORPORATED, Plaintiff-Counter-Defendant-Appellee, v. PLAYWOOD TOYS, INCORPORATED, Defendant-Counter-Plaintiff-Appellant, v. ROY WILSON, HARRY ABRAHAM, and JOHN LEE, Counter-Defendants-Appellees.
No. 02-1916
UNITED STATES COURT OF APPEALS FOR THE SEVENTH CIRCUIT
342 F.3d 714; 2003 U.S. App. LEXIS 16847; 67 U.S.P.Q.2D (BNA) 1801
December 13, 2002, Argued
August 18, 2003, Decided

PRIOR HISTORY: [**1] Appeal from the United States District Court for the Northern District of Illinois, Eastern Division. No. 94 C 6884. Rebecca R. Pallmeyer, Judge. PlayWood Toys, Inc. v. Learning Curve Toys, L.P., 2002 U.S. Dist. LEXIS 4298 (N.D. Ill., Mar. 14, 2002)

DISPOSITION: Reversed and remanded. Jury verdict reinstated.

COUNSEL: For LEARNING CURVE TOYS, INCORPORATED, Plaintiff - Appellee: Dean A. Dickie, ROOKS, PITTS & POUST, Chicago, IL USA. Roger L. Price, D'ANCONA & PFLAUM, Chicago, IL USA.

For PLAYWOOD TOYS, INCORPORATED, Defendant - Appellant: John S. Letchinger, WILDMAN, HARROLD, ALLEN & DIXON, Chicago, IL USA.

For ROY WILSON, HARRY ABRAHAM, JOHN LEE, Appellees: Dean A. Dickie, ROOKS, PITTS & POUST, Chicago, IL USA. Roger L. Price, D'ANCONA & PFLAUM, Chicago, IL USA.

JUDGES: Before RIPPLE, KANNE and ROVNER, Circuit Judges.

OPINIONBY: RIPPLE

OPINION:

[*716] RIPPLE, *Circuit Judge.* PlayWood Toys, Inc. ("PlayWood") obtained a jury verdict against Learning Curve Toys, Inc. and its representatives, Roy Wilson, Harry Abraham and John Lee (collectively, "Learning Curve"), for misappropriation of a trade secret in a realistic looking and sounding toy railroad track under the Illinois Trade Secrets Act, 765 ILCS 1065/1 *et seq*. The jury awarded PlayWood a royalty of "8% for a license that would have been negotiated [absent the misappropriation] to last for the lifetime of the product." R.194. Although there was substantial evidence of misappropriation before the jury, the district court did not enter judgment on the jury's verdict. Instead, it granted judgment as a matter of law in favor of Learning Curve, holding that PlayWood did not have a protectable trade secret in the toy railroad track. PlayWood appealed. For the reasons set forth in the following opinion, we reverse the judgment of the district court and reinstate the [**2] jury's verdict. We further remand the case to the district court for a jury trial on exemplary damages and for consideration of PlayWood's request for attorneys' fees.

I

BACKGROUND

A. Facts

In 1992, Robert Clausi and his brother-in-law, Scott Moore, began creating prototypes of wooden toys under the name PlayWood Toys,

Inc., a Canadian corporation. Clausi was the sole toy designer and Moore was the sole officer and director of PlayWood. Neither Clausi nor Moore had prior experience in the toy industry, but Clausi had "always been a bit of a doodler and designer," Trial Tr. at 58, and the two men desired to "create high-quality hardwood maple toys for the independent toy market," *id.* at 241. As a newly formed corporation, PlayWood did not own a facility in which it could produce toys. Instead, it worked in conjunction with Mario Borsato, who owned a woodworking facility. Subject to a written confidentiality agreement with PlayWood, Borsato manufactured prototypes for PlayWood based on Clausi's design specifications.

PlayWood's first attempt to market publicly its toys was at the Toronto Toy Fair on January 31, 1992. PlayWood received favorable reviews from many [**3] of the toy retailers in attendance; PlayWood also learned that the best way to get recognition for its toys was to attend the New York Toy Fair ("Toy Fair") the following month. Based on this information, Clausi and Moore secured a position at the Toy Fair in order to display PlayWood's prototypes. It was during this Toy Fair that Clausi and Moore first encountered Learning Curve representatives Roy Wilson, Harry Abraham and John Lee.

On the morning of February 12, 1993, the first day of the Toy Fair, Roy Wilson stopped at PlayWood's booth and engaged Clausi and Moore in conversation. Wilson identified himself as Learning Curve's toy designer and explained that his company had a license from the Britt Allcroft Company to develop Thomas the Tank Engine & Friends TM (hereinafter "Thomas") trains and accessories. Wilson commented that he was impressed with the look and quality of PlayWood's prototypes and raised the possibility of working together under a custom manufacturing contract to produce Learning Curve's line of Thomas products. Clausi and Moore responded that such an arrangement would be of great interest to PlayWood. Later that same day, Harry Abraham, Learning Curve's vice [**4] president, and John Lee, Learning Curve's president, also stopped by PlayWood's booth. They too commented [*717] on the quality of PlayWood's prototypes and indicated that PlayWood might be a good candidate for a manufacturing contract with Learning Curve.

Clausi and Moore continued to have discussions with Learning

Curve's representatives over the remaining days of the Toy Fair, which ended on February 14. During these discussions, Lee indicated that he would like two of his people, Abraham and Wilson, to visit PlayWood in Toronto the day after the Toy Fair ended in order to determine whether the two parties could work out a manufacturing arrangement for some or all of Learning Curve's wooden toys. Clausi, feeling a little overwhelmed by the suggestion, requested that their visit be postponed a few days so that he could better acquaint himself with Learning Curve's products. The parties ultimately agreed that Abraham and Wilson would visit PlayWood at Borsato's facility on February 18, 1993, four days after the conclusion of the Toy Fair. Clausi spent the next several days after the Toy Fair researching Learning Curve's products and considering how PlayWood could produce Learning Curve's [**5] trains and track.

On February 18, 1993, Abraham and Wilson visited PlayWood in Toronto as planned. The meeting began with a tour of Borsato's woodworking facility, where the prototypes on display at the Toy Fair had been made. After the tour, the parties went to the conference room at Borsato's facility. At this point, according to Clausi and Moore, the parties agreed to make their ensuing discussion confidential. Clausi testified:

> After we sat down in the board room, Harry [Abraham of Learning Curve] immediately said: "Look, we're going to disclose confidential information to you guys, and we're going to disclose some designs that Roy [Wilson of Learning Curve] has that are pretty confidential. If Brio were to get their hands on them, then we wouldn't like that. And we're going to do it under the basis of a confidential understanding."
>
> And I said: "I also have some things, some ideas on how to produce the track and produce the trains now that I've had a chance to look at them for the last couple of days, and I think they're confidential as well. So if we're both okay with that, we should continue." So we did.

Trial Tr. at 76-77. Moore testified to the existence of [**6] a similar conversation:

> It was at this point that Harry Abraham told us that they were going to disclose some confidential documents, drawings, pricing, margins, and asked us if we would keep that information confidential.
>
> * * * *
>
> I believe it was Robert [Clausi] who said that, you know, absolutely, we would keep it confidential. In fact, we had some ideas that we felt would be confidential we would be disclosing to them, and would they keep it, you know, confidential? Would they reciprocate? And Harry [Abraham] said: "Absolutely." And then we proceeded to go along with the meeting.

Trial Tr. at 247-48.

Immediately after the parties agreed to keep their discussion confidential, Wilson, at Abraham's direction, showed Clausi and Moore drawings of various Thomas characters and provided information on the projected volume of each of the products. Clausi testified that he considered the documents disclosed by Learning Curve during the meeting confidential because they included information on products not yet released to the public, as well as Learning Curve's projected volumes, costs and profit margins for various products. After viewing [*718] Wilson's various drawings, the [**7] parties discussed PlayWood's ideas on how to manufacture Learning Curve's trains. Clausi suggested that they might use a CNC machine, which he defined as a computer numerically controlled drill that carves in three dimensions, to create Learning Curve's trains out of a single piece of wood (as opposed to piecing together separate pieces of wood).

The parties' discussion eventually moved away from train production and focused on track design. Wilson showed Clausi and Moore drawings of Learning Curve's track and provided samples of their current product. At this point, Abraham confided to Clausi and Moore that track had posed "a bit of a problem for Learning Curve." Trial Tr. at 85. Abraham explained that sales were terrific for

Learning Curve's Thomas trains, but that sales were abysmal for its track. Abraham attributed the lack of sales to the fact that Learning Curve's track was virtually identical to that of its competitor, Brio, which had the lion's share of the track market. Because there was "no differentiation" between the two brands of track, Learning Curve's track was not even displayed in many of the toy stores that carried Learning Curve's products. *Id.* Learning Curve [**8] had worked unsuccessfully for several months attempting to differentiate its track from that of Brio.

After detailing the problems with Learning Curve's existing track, Abraham inquired of Clausi whether "there was a way to differentiate" its track from Brio's track. Trial Tr. at 86. Clausi immediately responded that he "had had a chance to look at the track and get a feel for it [over] the last few days" and that his "thoughts were that if the track were more realistic and more functional, that kids would enjoy playing with it more and it would give the retailer a reason to carry the product, especially if it looked different than the Brio track." *Id.* at 87. Clausi further explained that, if the track "made noise and [] looked like real train tracks, that the stores wouldn't have any problem, and the Thomas the Tank line, product line would have [] its own different track" and could "effectively compete with Brio." *Id.* Abraham and Wilson indicated that they were "intrigued" by Clausi's idea and asked him what he meant by "making noise." *Id.*

Clausi decided to show Abraham and Wilson exactly what he meant. Clausi took a piece of Learning Curve's existing track from [**9] the table, drew some lines across the track (about every three-quarters of an inch), and stated: "We can go ahead and machine grooves right across the upper section . . ., which would look like railway tracks, and down below machine little indentations as well so that it would look more like or sound more like real track. You would roll along and bumpity-bumpity as you go along." Trial Tr. at 255. Clausi then called Borsato into the conference room and asked him to cut grooves into the wood "about a quarter of an inch deep from the top surface." *Id.* at 88. Borsato left the room, complied with Clausi's request, and returned with the cut track three or four minutes later. Clausi ran a train back and forth over the cut piece of track. The track looked more realistic than before, but it did not make noise because the grooves were not deep enough. Accordingly, Clausi instructed Borsato to cut the grooves "just a little bit deeper so that they go through the rails." *Id.* Borsato complied with Clausi's request once again and returned a few minutes later with the cut piece of track. Clausi proceeded to run a train back and forth over the track. This time the track made a "clickety-clack" [**10] sound, but the train did not run smoothly over the track because the grooves were cut "a little bit too deep." [*719] *Id.* at 258. Based on the sound produced by the track, Clausi told Abraham and Moore that if PlayWood procured a contract with Learning Curve to produce the track, they could call it "Clickety-Clack Track." *Id.* at 89.

Both Abraham and Wilson indicated that Clausi's concept of cutting grooves into the track to produce a clacking sound was a novel concept. Thereafter, Wilson and Clausi began to discuss how they could improve the idea to make the train run more smoothly on the track, but Abraham interrupted them and stated: "No, focus. You guys have to get the contract for the basic product first, and then we can talk about new products, because . . . it takes [our licensor] a long time to approve new products and new designs." Trial Tr. at 89.

The meeting ended shortly thereafter without further discussion about Clausi's concept for the noise-producing track. Before he left, Wilson asked Clausi if he could take the piece of track that Borsato had cut with him while the parties continued their discussions. Clausi gave Wilson the piece of track without hesitation. The [**11] piece of track was the only item that Abraham and Wilson took from the meeting. Clausi and Moore did not ask Wilson for a receipt for the cut track, nor did they seek a written confidentiality agreement to protect PlayWood's alleged trade secret. After the meeting, Clausi amended PlayWood's confidentiality agreement with Borsato to ensure that materials discussed during the meeting would remain confidential. Clausi also stamped many of the documents that he received from Learning Curve during the meeting as confidential because they included information on products not yet released to the public. PlayWood never disclosed the contents of Learning Curve's documents to anyone.

During March of 1993, PlayWood and Learning Curve met on three separate occasions to discuss further the possibility of

PlayWood manufacturing Learning Curve's Thomas products. At one of the meetings, and at Learning Curve's request, PlayWood submitted a manufacturing proposal for the Thomas products. Learning Curve rejected PlayWood's proposal. Learning Curve told Clausi that its licensor wanted the Thomas products to be made in the United States.

Thereafter, PlayWood had no contact with Learning Curve until [**12] late October of 1993, when Abraham contacted Clausi to discuss another possible manufacturing contract because Learning Curve's secondary supplier was not providing enough product. Again, PlayWood submitted a manufacturing proposal at Learning Curve's request, but it too was rejected. Learning Curve later stated that its new business partner had decided to manufacture the product in China.

Clausi and Moore continued to work on PlayWood's toy concepts. After the 1994 New York Toy Fair, which was not particularly successful for PlayWood, Clausi and Moore began to focus their efforts on refining PlayWood's concept for the noise-producing track. During this time, Clausi and Moore made no attempt to license or sell the concept to other toy companies because they believed that PlayWood still had "an opportunity to get in the door" with Learning Curve if they could perfect the concept and also because they believed that they were bound by a confidentiality agreement. Trial Tr. at 267.

In December of 1994, while shopping for additional track with which to experiment, Moore discovered that Learning Curve was selling noise-producing track under the name "Clickety-Clack Track." Like the piece [**13] of track that Clausi had Borsato cut during PlayWood's February 18, 1993, [*720] meeting with Learning Curve, Clickety-Clack Track TM has parallel grooves cut into the wood, which cause a "clacking" sound as train wheels roll over the grooves. Learning Curve was promoting the new track as

> the first significant innovation in track design since the inception of wooden train systems. . . . It is quite simply the newest and most exciting development to come along recently in the wooden train industry, and it's sure to

> cause a sensation in the marketplace. . . . It brings that sound and feel of the real thing to a child's world of make-believe without bells, whistles, electronic sound chips or moving parts.

PlayWood's Tr. Ex.71.

Moore was "stunned" when he saw the track because he believed that Learning Curve had stolen PlayWood's concept. Trial Tr. at 268. He testified: "This was our idea. This is what we've been working on even up to that day to go back to [Learning Curve] as an opportunity to get in the door, and there it is on the shelf." *Id.* Moore purchased a package of Clickety-Clack Track TM and showed it to Clausi. Clausi testified that he was disappointed when [**14] he saw the track because he believed that Learning Curve had taken PlayWood's name and design concept "almost exactly as per [their] conversation" on February 18, 1993. Trial Tr. at 103.

PlayWood promptly wrote a cease and desist letter to Learning Curve. The letter accused Learning Curve of stealing PlayWood's concept for the noise-producing track that it disclosed to Learning Curve "in confidence in the context of a manufacturing proposal." PlayWood's Tr. Ex. 66 at 1. Learning Curve responded by seeking a declaratory judgment that it owned the concept.

Previously, on March 16, 1994, Learning Curve had applied for a patent on the noise-producing track. The patent, which was obtained on October 3, 1995, claims the addition of parallel impressions or grooves in the rails, which cause a "clacking" sound to be emitted as train wheels roll over them. The patent identifies Roy Wilson of Learning Curve as the inventor.

Clickety-Clack Track TM provided an enormous boost to Learning Curve's sales. Learning Curve had $ 20 million in track sales by the first quarter of 2000, and $ 40 million for combined track and accessory sales.

B. District Court Proceedings

Learning Curve [**15] responded to PlayWood's cease and desist letter by seeking a declaratory judgment that it owned the concept for noise-producing toy railroad track, as embodied in Clickety-Clack Track. TM PlayWood counterclaimed against Learning Curve, as well as its representatives, Roy Wilson, Harry Abraham and John Lee. PlayWood asserted that it owned the concept and that Learning Curve had misappropriated its trade secret. n1 Learning Curve voluntarily dismissed its complaint for declaratory relief, [*721] and PlayWood's claim for trade secret misappropriation proceeded to trial. The jury returned a verdict in favor of PlayWood. The trial court declined to enter judgment on the verdict and instead asked the parties to brief Learning Curve's Rule 50 motion on the issue of whether PlayWood had a protectable trade secret under the Illinois Trade Secrets Act, 765 ILCS 1065/1 *et seq*. The district court granted Learning Curve's motion and entered judgment in its favor on the ground that PlayWood presented insufficient evidence of a trade secret. *See* R.202. Specifically, the court determined that PlayWood did not have a trade secret in its concept for noise-producing toy railroad track under Illinois [**16] law because: (1) PlayWood did not demonstrate that its concept was unknown in the industry; (2) PlayWood's concept could have been easily acquired or duplicated through proper means; (3) PlayWood failed to guard the secrecy of its concept; (4) PlayWood's concept had no economic value; and (5) PlayWood expended no time, effort or money to develop the concept. *See id.*

n1 In its amended counterclaim, PlayWood asserted eight causes of action: (1) implied-in-fact contract; (2) quasi-contract; (3) idea misappropriation; (4) fraud and deceptive business practices under the Illinois Consumer Fraud and Deceptive Business Practices Act, 815 ILCS 505/1 *et seq*.; (5) misappropriation of trade secrets under the Illinois Trade Secrets Act, 765 ILCS 1065/1 *et seq*.; (6) unfair competition under § 44(b) of the Lanham Act, 15 U.S.C. § 1126; (7) unfair competition under § 43(a) of the Lanham Act, 15 U.S.C. § 1125; and (8) deceptive trade practices under the Uniform Deceptive Trade Practices Act, 815 ILCS § 510/1 *et seq*. R.35. With the exception of misappropriation of trade secrets, the district court entered summary judgment against PlayWood on all counts. PlayWood does not appeal the grant of summary judgment on those counts.

[**17]
II
DISCUSSION

A. Trade Secret Status

We review the district court's decision to grant Learning Curve's motion for judgment as a matter of law de novo, viewing the evidence in the light most favorable to Play-Wood. *See Veach v. Sheeks*, 316 F.3d 690, 692 (7th Cir. 2003). "We shall not second-guess the jury's view of the contested evidence; the proper inquiry is whether, given the totality of the evidence, [PlayWood] presented sufficient evidence from which a reasonable jury could find in [its] favor." *David v. Caterpillar, Inc.*, 324 F.3d 851, 858 (7th Cir. 2003).

The parties agree that their dispute is governed by the Illinois Trade Secrets Act ("Act"), 765 ILCS 1065/1 *et seq*. To prevail on a claim for misappropriation of a trade secret under the Act, the plaintiff must demonstrate that the information at issue was a trade secret, that it was misappropriated and that it was used in the defendant's business. *See Composite Marine Propellers, Inc. v. Van Der Woude,* 962 F.2d 1263, 1265-66 (7th Cir. 1992) (per curiam); *Southwest Whey, Inc. v. Nutrition 101, Inc.*, 117 F. Supp. 2d 770, 775-76 (N.D. Ill. 2000); [**18] *Magellan Int'l. Corp. v. Salzgitter Handel GmbH.,* 76 F. Supp. 2d 919, 926 (N.D. Ill. 1999). The issue currently before us is whether there was legally sufficient evidence for the jury to find that PlayWood had a trade secret in its concept for the noise-producing toy railroad track that it revealed to Learning Curve on February 18, 1993.

The Act defines a trade secret as:

> Information, including but not limited to, technical or non-technical data, a formula, pattern, compilation, program, device, method, technique, drawing, process, financial data, or list of actual or potential customers or suppliers, that:

> (1) is sufficiently secret to derive economic value, actual or potential, from not being generally known to other persons who can obtain economic value from its disclosure or use; and
>
> (2) is the subject of efforts that are reasonable under the circumstances to maintain its secrecy or confidentiality.

765 ILCS 1065/2(d). Both of the Act's statutory requirements focus fundamentally on the secrecy of the information sought to be protected. *See Mangren Research & Dev. Corp. v. Nat'l Chem. Co., Inc.*, 87 F.3d 937, 942 (7th Cir. 1996); [**19] *Computer Care v. Serv. Sys. Enters., Inc.*, 982 F.2d 1063, 1072 (7th Cir. 1992); *Pope v. Alberto-Culver* [*722] *Co.*, 296 Ill. App. 3d 512, 694 N.E.2d 615, 617, 230 Ill. Dec. 646 (Ill. App. Ct. 1998); *Stampede Tool Warehouse, Inc. v. May*, 272 Ill. App. 3d 580, 651 N.E.2d 209, 215, 209 Ill. Dec. 281 (Ill. App. Ct. 1995); *Serv. Ctrs. of Chicago, Inc. v. Minogue*, 180 Ill. App. 3d 447, 535 N.E.2d 1132, 1136, 129 Ill. Dec. 367 (Ill App. Ct. 1989). However, the requirements emphasize different aspects of secrecy. The first requirement, that the information be sufficiently secret to impart economic value because of its relative secrecy, "precludes trade secret protection for information generally known or understood within an industry even if not to the public at large." *Pope*, 694 N.E.2d at 617. The second requirement, that the plaintiff take reasonable efforts to maintain the secrecy of the information, prevents a plaintiff who takes no affirmative measures to prevent others from using its proprietary information from obtaining trade secret protection. *See Jackson v. Hammer*, 274 Ill. App. 3d 59, 653 N.E.2d 809, 816, 210 Ill. Dec. 614 (Ill. App. Ct. 1995) [**20] ("The Act requires a plaintiff to take 'affirmative measures' to prevent others from using information.").

Although the Act explicitly defines a trade secret in terms of these two requirements, Illinois courts frequently refer to six common law factors (which are derived from § 757 of the Restatement (First) of Torts) in determining whether a trade secret exists: (1) the extent to which the information is known outside of the plaintiff's business; (2) the extent to which the information is known by employees and others involved in the plaintiff's business; (3) the extent of measures taken by the plaintiff to guard the secrecy of the information; (4) the value of the information to the plaintiff's business and to its competitors; (5) the amount of time, effort and money expended by the plaintiff in developing the information; and (6) the ease or difficulty with which the information could be properly acquired or duplicated by others. *See Delta Med. Sys. v. Mid-America Med. Sys., Inc.,* 331 Ill. App. 3d 777, 772 N.E.2d 768, 780, 265 Ill. Dec. 397 (Ill. App. Ct. 2002); *Stampede Tool Warehouse,* 651 N.E.2d at 215-16; *George S. May Int'l Co. v. Int'l Profit Assocs.*, 256 Ill. App. 3d 779, 628 N.E.2d 647, 653, 195 Ill. Dec. 183 (Ill. App. Ct. 1993); [**21] *see also C&F Packing Co., Inc. v. IBP, Inc.,* 224 F.3d 1296, 1302 (Fed. Cir. 2000) (applying Illinois law).

Contrary to Learning Curve's contention, we do not construe the foregoing factors as a six-part test, in which the absence of evidence on any single factor necessarily precludes a finding of trade secret protection. Instead, we interpret the common law factors as instructive guidelines for ascertaining whether a trade secret exists under the Act. The language of the Act itself makes no reference to these factors as independent requirements for trade secret status, and Illinois case law imposes no such requirement that each factor weigh in favor of the plaintiff. *See ILG Indus., Inc. v. Scott*, 49 Ill. 2d 88, 273 N.E.2d 393, 396 (Ill. 1971) ("An exact definition of a trade secret, applicable to all situations, is not possible. Some factors to be considered in determining whether given information is one's trade secret are [the six factors enumerated in the Restatement].") (internal quotation marks omitted). In this respect, Illinois law is compatible with the approach in other states. Courts from other jurisdictions, as well as legal scholars, [**22] have noted that the Restatement factors are not to be applied as a list of requisite elements. *See, e.g., Basic American, Inc. v. Shatila*, 133 Idaho 726, 992 P.2d 175, 184 (Idaho 1999); *Minuteman, Inc. v. Alexander*, 147 Wis. 2d 842, 434 N.W.2d 773, 778

(Wis. 1989); 2 Gregory E. Upchurch, *Intellectual Property Litigation Guide: Patents & Trade Secrets* § 16.02, [*723] at 16-17 to 16-18 (2002) ("On the whole, these factors are a guide to the proper decision on the existence of a trade secret, not a list of requirements.").

The existence of a trade secret ordinarily is a question of fact. n2 As aptly observed by our colleagues on the Fifth Circuit, a trade secret "is one of the most elusive and difficult concepts in the law to define." *Lear Siegler, Inc. v. Ark-Ell Springs, Inc.,* 569 F.2d 286, 288 (5th Cir. 1978). In many cases, the existence of a trade secret is not obvious; it requires an ad hoc evaluation of all the surrounding circumstances. For this reason, the question of whether certain information constitutes a trade secret ordinarily is best "resolved by a fact finder after full presentation of evidence from each side." *Id.* at 289. [**23] We do not believe that the district court was sufficiently mindful of these principles. The district court, in effect, treated the Restatement factors as requisite elements and substituted its judgment for that of the jury. PlayWood presented sufficient evidence for the jury reasonably to conclude that the Restatement factors weighed in PlayWood's favor.

n2 *See Nilssen v. Motorola, Inc.,* 963 F. Supp. 664, 675 (N.D. Ill. 1997) (applying Illinois law); *see also Penalty Kick Mgmt. Ltd. v. Coca Cola Co.*, 318 F.3d 1284, 1291 (11th Cir. 2003) (applying Georgia law); *Pate v. Nat'l Fund Raising Consultants, Inc.,* 20 F.3d 341, 344 (8th Cir. 1994) (applying Colorado law); *Chevron U.S.A. Inc. v. Roxen Serv., Inc.,* 813 F.2d 26, 29 (2d Cir. 1987) (applying New York law); 1 Melvin F. Jager, *Trade Secrets Law* § 5.2, at 5-3 (2002); 2 Gregory E. Upchurch, *Intellectual Property Litigation Guide: Patents & Trade Secrets* § 16.03, at 16-18 (2002).

1. [**24] **Extent to which PlayWood's concept for noise-producing toy railroad track was known outside of Playwood's business**

PlayWood presented substantial evidence from which the jury could have determined that PlayWood's concept for noise-producing toy railroad track was not generally known outside of Playwood's business. It was undisputed at trial that no similar track was on the market until Learning Curve launched Clickety-Clack Track TM in late 1994, more than a year after PlayWood first conceived of the concept. Of course, as Learning Curve correctly points out, "merely being the first or only one to use particular information does not in and of itself transform otherwise general knowledge into a trade secret." *George S. May Int'l*, 628 N.E.2d at 654. "If it did, the first person to use the information, no matter how ordinary or well known, would be able to appropriate it to his own use under the guise of a trade secret." *Serv. Ctrs.*, 535 N.E.2d at 1137. However, in this case, there was additional evidence from which the jury could have determined that PlayWood's concept was not generally known within the industry.

First, there was substantial testimony [**25] that Learning Curve had attempted to differentiate its track from that of its competitors for several months, but that it had been unable to do so successfully.

Furthermore, PlayWood's expert witness, Michael Kennedy, testified that PlayWood's concept, as embodied in Clickety-Clack Track TM, was unique and permitted "its seller to differentiate itself from a host of competitors who [were] making a generic product." Trial Tr. at 518. Kennedy explained that the look, sound and feel of the track made it distinct from other toy railroad track: "When a child runs a train across this track, he can feel it hitting those little impressions. And when you're talking about young children[,] having the idea that they can see something [*724] that they couldn't see before, feel something that they couldn't feel before, hear something that they couldn't hear before, that is what differentiates this toy from its other competitors." *Id.* at 489.

Finally, PlayWood presented evidence that Learning Curve sought and obtained a patent on the noise-producing track. It goes without saying that the requirements for patent and trade secret protection are not synonymous. Unlike "a patentable [**26] invention, a trade secret need not be novel or unobvious." 2 Rudolf Callmann, *The Law of Unfair Competition, Trademarks and Monopolies* § 14.15, at 14-124 (4th ed. 2003). "The idea need not be complicated; it may be intrinsically simple and nevertheless qualify as a secret, unless it is common knowledge and, therefore, within the public domain." *Forest*

Labs, Inc. v. Pillsbury Co., 452 F.2d 621, 624 (7th Cir. 1971) (internal quotation marks omitted). However, it is commonly understood that "if an invention has sufficient novelty to be entitled to patent protection, it may be said *a fortiori* to be entitled to protection as a trade secret." 1 Roger M. Milgrim, *Milgrim on Trade Secrets* § 1.08[1], at 1-353 (2002) (internal footnotes omitted). In light of this evidence, we cannot accept Learning Curve's argument that no rational jury could have found that PlayWood's concept was unknown outside of its business.

2. Extent to which PlayWood's concept was known to employees and others involved in PlayWood's business

The district court did not address the extent to which PlayWood's concept was known to employees and others involved in PlayWood's business. [**27] However, we agree with PlayWood that the evidence was sufficient to establish that its concept for noise-producing track was known only by key individuals in its business.

At the outset, we note briefly that PlayWood was a small business, consisting only of Clausi and Moore. Illinois courts have recognized on several occasions that the expectations for ensuring secrecy are different for small companies than for large companies. *See Jackson*, 653 N.E.2d at 815 ("The determination of what steps are reasonably necessary to protect information is different for a large company than for a small one."); *Elmer Miller, Inc. v. Landis*, 253 Ill. App. 3d 129, 625 N.E.2d 338, 342, 192 Ill. Dec. 378 (Ill. App. Ct. 1993) ("Reasonable steps for a two or three person shop may be different from reasonable steps for a larger company."). Apart from Clausi (PlayWood's sole toy designer and the person who conceived of the concept for noise-producing track) and Moore (PlayWood's sole officer and director), the only person who knew about the concept was Borsato, the person who physically produced PlayWood's prototype at Clausi's direction. The concept was disclosed to Borsato [**28] in order for PlayWood to develop fully its trade secret. *See* 1 Roger M. Milgrim, *Milgrim on Trade Secrets* § 1.04, at 1-173 (2002) ("A trade secret does not lose its character by being confidentially disclosed to agents or servants, without whose assistance it could not be made of any value.") (internal quotation marks omitted). Moreover, Borsato's actions were governed by a written confidentiality agreement with PlayWood. Indeed, as an extra precaution, Clausi even amended PlayWood's confidentiality agreement with Borsato immediately after the February 18, 1993, meeting to ensure that materials discussed during the meeting would remain confidential. From this evidence, the jury reasonably could have determined that this factor also weighed in favor of PlayWood. [*725]

3. Measures taken by PlayWood to guard the secrecy of its concept

There also was sufficient evidence for the jury to determine that PlayWood took reasonable precautions to guard the secrecy of its concept. The Act requires the trade secret owner to take actions that are "reasonable under the circumstances to maintain [the] secrecy or confidentiality" of its trade secret; it does not require perfection. [**29] 765 ILCS 1065/2(d)(2). Whether the measures taken by a trade secret owner are sufficient to satisfy the Act's reasonableness standard ordinarily is a question of fact for the jury. n3 Indeed, we previously have recognized that "only in an extreme case can what is a 'reasonable' precaution be determined [as a matter of law], because the answer depends on a balancing of costs and benefits that will vary from case to case." *Rockwell Graphic Sys., Inc. v. DEV Indus., Inc.*, 925 F.2d 174, 179 (7th Cir. 1991).

n3 *See Mangren Research & Dev. Corp. v. Nat'l Chem. Co., Inc.*, 87 F.3d 937, 943 (7th Cir. 1996); *Rockwell Graphic Sys., Inc. v. DEV Indus., Inc.*, 925 F.2d 174, 179 (7th Cir. 1991); *see also* 1 Roger M. Milgrim, *Milgrim on Trade Secrets* § 1.04, at 1-170 (2002); 2 Rudolf Callmann, *The Law of Unfair Competition, Trademarks and Monopolies* § 14.26, at 14-209 (4th ed. 2003).

Here, the jury was instructed that it must find "by a preponderance of the evidence [**30] that PlayWood's trade secrets were given to Learning Curve as a result of a confidential relationship between the parties." Trial Tr. at 1449. By returning a verdict in favor of PlayWood, the jury necessarily found that Learning Curve was bound to PlayWood by a pledge of confidentiality. The jury's determination is amply supported by the evidence. Both Clausi and Moore testified that they entered into an oral confidentiality agreement with Abraham and Wilson before beginning their discussion on February 18, 1993. In particular, Clausi testified that he told Abraham and Wilson: "I also have some things, some ideas on how to produce the track and produce the trains now that I've had a chance to look at them for the last couple of days, and I think they're confidential as well. So if we're both okay with that, we should continue." Trial Tr. at 77. In addition to this testimony, the jury heard that Learning Curve had disclosed substantial information to PlayWood during the February 18th meeting, including projected volumes, costs and profit margins for various products, as well as drawings for toys not yet released to the public. The jury could have inferred that Learning Curve would not [**31] have disclosed such information in the absence of a confidentiality agreement. Finally, the jury also heard (from several of Learning Curve's former business associates) that Learning Curve routinely entered into oral confidentiality agreements like the one with PlayWood.

PlayWood might have done more to protect its secret. As Learning Curve points out, PlayWood gave its only prototype of the noise-producing track to Wilson without first obtaining a receipt or written confidentiality agreement from Learning Curve—a decision that proved unwise in hindsight. Nevertheless, we believe that the jury was entitled to conclude that PlayWood's reliance on the oral confidentiality agreement was reasonable under the circumstances of this case. n4 First, it is well [*726] established that "the formation of a confidential relationship imposes upon the disclosee the duty to maintain the information received in the utmost secrecy" and that "the unprivileged use or disclosure of another's trade secret becomes the basis for an action in tort." *Burten v. Milton Bradley Co.*, 763 F.2d 461, 463 (1st Cir. 1985). Second, both Clausi and Moore testified that they believed PlayWood had a realistic [**32] chance to "get in the door" with Learning Curve and to produce the concept as part of Learning Curve's line of Thomas products. Clausi and Moore did not anticipate that Learning Curve would violate the oral confidentiality agreement and utilize PlayWood's concept without permission; rather, they believed in good faith that they "were going to do business one day again with Learning Curve with respect to the design concept." Trial Tr. at 236-37. Finally, we believe that, as part of the reasonableness inquiry, the jury could have considered the size and sophistication of the parties, as well as the relevant industry. Both PlayWood and Learning Curve were small toy companies, and PlayWood was the smaller and less experienced of the two. Viewing the evidence in the light most favorable to PlayWood, as we must, we conclude that there was sufficient evidence for the jury to determine that PlayWood took reasonable measures to protect the secrecy of its concept.

> n4 We iterate that the proper inquiry is not whether, in our independent judgment, we believe that PlayWood took reasonable precautions to maintain the secrecy of its concept; rather, the issue is whether PlayWood's "failure to do more was so plain a breach of the obligation of a trade secret owner to make reasonable efforts to maintain secrecy as to justify" overturning the jury verdict in its favor. *Rockwell*, 925 F.2d at 177.

[**33]

4. Value of the concept to PlayWood and to its competitors

There was substantial evidence from which the jury could have determined that PlayWood's concept had value both to PlayWood and to its competitors. It was undisputed at trial that Learning Curve's sales skyrocketed after it began to sell Clickety-Clack Track TM. In addition, PlayWood's expert witness, Michael Kennedy, testified that PlayWood's concept for noise-producing track had tremendous value. Kennedy testified that the "cross-cuts and changes in the [track's] surface" imparted value to its seller by causing the track to "look different, feel different and sound different than generic track." Trial Tr. at 504. Kennedy further testified that, in his opinion, the track would have commanded a premium royalty under a negotiated license agreement because the "invention allows its seller to differentiate itself from a host of competitors who are making a generic product with whom it is competing in a way that is proprietary and exclusive, and it gives [the seller] a significant edge over [its] competition." *Id.*

at 518-19.

Despite this evidence, the district court concluded that PlayWood's concept had no economic [**34] value. The court's conclusion was based, in part, on the fact that PlayWood's prototype did not work perfectly; as noted by the court, the first set of cuts were too shallow to produce sound and the second set of cuts were too deep to permit the train to roll smoothly across the track. In the district court's view, even if the concept of cutting grooves into the wooden track in order to produce noise originated with Clausi, the concept lacked value until it was refined, developed and manufactured by Learning Curve.

We cannot accept the district court's conclusion because it is belied by the evidence. At trial, Kennedy was asked whether, in his opinion, the fact that PlayWood's prototype did not work perfectly affected the value of PlayWood's concept, and he testified that it did not. *See* Trial Tr. at 578. Kennedy testified that he would assign the same value to PlayWood's concept as it was conceived on February 18, 1993, as he would the finished product that became known as Clickety-Clack Track TM because, at that time, he would have known "that most of the [*727] design [had] already been done and that [he] just needed to go a little bit further to make it really lovely." *Id.* [**35] at 578. Kennedy further testified that it was standard practice in the industry for a license to be negotiated based on a prototype (much like the one PlayWood disclosed to Learning Curve) rather than a finished product and that the license generally would cover the prototypical design, as well as any enhancements or improvements of that design. *See* Trial Tr. at 500-01. n5 Based on this testimony, we cannot accept the district court's conclusion that Play-Wood's concept possessed no economic value.

n5 Specifically, Kennedy testified:

> Q: Now, when you were at Tyco, you were in the toy business and people were bringing you inventions, did people bring you inventions or trade secrets that were in the kind of form that we

find Defendants' Exhibit 9 [Clickety-Clack Track], fully polished and finished and so on?

A: I've seen some pretty rough-looking toys in meeting with inventors. I've seen toys that were obviously made by hand. I've seen toys that had cracks, seams and joints that you don't expect to see when they're manufactured. Certainly that's true. . . . So the answer is: You don't see a final product when you meet with an inventor. You see a preliminary product or a prototype kind of product.

Q: Now, when a prototype is brought to you as a disclosure, as a secret, as an invention that somebody wants to license to you, does it make a difference to you whether it's a prototype or a finished product?

A: Not necessarily, because it depends on who is going to make it, which is uncertain at the time. It depends on how difficult it is to make, which could be uncertain at that time. It's helpful if we know that it's easy to make. It's helpful if we know how much it costs to make. But you don't always know that.

Q: When you go to license that kind of invention that's brought to you, is it your intent to license only the prototype that's brought to you?

A: No. I think every license agreement that I negotiate in the toy industry includes the prototypical design. It includes enhancements and improvements on that design, regardless of whether they're made by the inventor or whether they're made by the manufacturer. It includes something called line extensions, which is the transfer of this invention to a toy which maybe wasn't first thought of for its application. It includes all of those things.

Trial Tr. at 499-501.

[**36]

It is irrelevant under Illinois law that PlayWood did not actually use the concept in its business. "The proper criterion is not 'actual use' but whether the trade secret is 'of value' to the company." *Syntex Ophthalmics, Inc. v. Tsuetaki*, 701 F.2d 677, 683 (7th Cir. 1983). n6 Kennedy's testimony was more than sufficient to permit the jury to conclude that the concept was "of value" to PlayWood. It is equally irrelevant that PlayWood did not seek to patent its concept. So long as the concept remains a secret, *i.e.*, outside of the public domain, there is no need for patent protection. Professor Milgrim makes this point well: "Since every inventor has the right to keep his invention secret, one who has made a patentable invention has the option to maintain it in secrecy, relying upon protection accorded to a trade secret rather [*728] than upon the rights which accrue by a patent grant." 1 Roger M. Milgrim, *Milgrim on Trade Secrets* § 1.08[1], at 1-353 (2002). It was up to PlayWood, not the district court, to determine when and how the concept should have been disclosed to the public.

n6 Both the Uniform Trade Secrets Act and the Restatement (Third) of Unfair Competition expressly reject prior use by the person asserting rights in the information as a prerequisite to trade secret protection. *See* Unif. Trade Secrets Act § 1 cmt. (1990) ("The broader definition in the proposed Act extends protection to a plaintiff who has not yet had an opportunity or acquired the means to put a trade secret to use."); Restatement (Third) of Unfair Competition § 39 cmt. e (1995) ("Use by the person asserting rights in the information is not a prerequisite to protection under the rule stated in this Section," in part, because such a "requirement can deny protection during periods of research and development and is particularly burdensome for innovators who do not possess the capability to exploit their innovations.").

[**37]

5. Amount of time, effort and money expended by PlayWood in developing its concept

PlayWood expended very little time and money developing its concept; by Clausi's own account, the cost to PlayWood was less than one dollar and the time spent was less than one-half hour. The district court determined that "such an insignificant investment is . . . insufficient as a matter of Illinois law to establish the status of a 'trade secret.' " R.202 at 16. We believe that the district court gave too much weight to the time, effort and expense of developing the track. n7

n7 Professor Milgrim, for one, rejects any per se requirement of developmental costs:

> Where cost is referred to it is almost always invariably incidental to other, basic definitional elements, such as secrecy. Since it is established that a trade secret can be discovered fortuitously (ergo, without costly development), or result purely from the exercise of creative facilities, it would appear inconsistent to consider expense of development of a trade secret as an operative substantive element.

See 1 Roger M. Milgrim, *Milgrim on Trade Secrets* § 1.02[2], at 1-146 & 1-150 (2002) (internal footnotes omitted).

[**38]

Although Illinois courts commonly look to the Restatement factors for guidance in determining whether a trade secret exists, as we have noted earlier, the requisite statutory inquiries under Illinois law are (1) whether the information "is sufficiently secret to derive economic value, actual or potential, from not being generally known to other persons who can obtain economic value from its disclosure or use;" and (2) whether the information "is the subject of efforts that are reasonable under the circumstances to maintain its secrecy or confidentiality." 765 ILCS 1065/2(d). A significant expenditure of time and/or money in the production of information may provide evidence of value, which is relevant to the first inquiry above. However, we do not understand Illinois law to require such an expenditure in all cases.

As pointed out by the district court, several Illinois cases have emphasized the importance of developmental costs. However, notably, none of those cases concerned the sort of innovative and creative concept that we have in this case. Indeed, several of the cases in Illinois that emphasize developmental costs concern compilations of data, such as customer lists. n8 In [**39] that context, it makes sense to require the [*729] expenditure of significant time and money because there is nothing original or creative about the alleged trade secret. Given enough time and money, we presume that the plaintiff's competitors could compile a similar list.

n8 *See, e.g., Delta Med. Sys. v. Mid-America Med. Sys., Inc.*, 331 Ill. App. 3d 777, 772 N.E.2d 768, 781, 265 Ill. Dec. 397 (Ill. App. Ct. 2002) ("Delta presented no testimony at the hearing as to the amount of effort expended in acquiring its customer list."); *Strata Mktg., Inc. v. Murphy,* 317 Ill. App. 3d 1054, 740 N.E.2d 1166, 1177, 251 Ill. Dec. 595 (Ill. App. Ct. 2000) ("Strata's customer lists, which it alleged take considerable effort, time, and money to compile, could be deemed a trade secret."); *Stampede Tool Warehouse, Inc. v. May*, 272 Ill. App. 3d 580, 651 N.E.2d 209, 216, 209 Ill. Dec. 281 (Ill. App. Ct. 1995) ("The customer list has been developed through the laborious method of prospecting, which requires a substantial amount of time, effort, and expense by Stampede."); *Springfield Rare Coin Galleries, Inc. v. Mileham*, 250 Ill. App. 3d 922, 620 N.E.2d 479, 485, 189 Ill. Dec. 511 (Ill. App. Ct. 1993) ("Under Illinois law, customer lists and other customer information will constitute confidential information *only* when the information has been developed by the employer over a number of years at great expense and kept under tight security."); *Abbott-Interfast Corp. v. Harkabus*, 250 Ill. App. 3d 13, 619 N.E.2d 1337, 1344, 189 Ill. Dec. 288 (Ill. App. Ct. 1993) ("Items such as customer lists, pricing information, and business techniques can be trade secrets if the employer has developed the information over a number of years at great expense and kept [it] under tight security.") (internal quotation marks omitted); *Prudential Ins. Co. of America v. Van Matre*, 158 Ill. App. 3d 298, 511 N.E.2d 740, 745, 110 Ill. Dec. 563 (Ill. App. Ct. 1987) ("A customer list or other customer information constitutes a trade secret in which an employer holds a protectable interest where the employer developed the information over a number of years, at great expense, and kept the information under lock and key.").

[**40]

Here, by contrast, we are dealing with a new toy design that has been promoted as "the first significant innovation in track design since the inception of wooden train systems." PlayWood's Tr. Ex.71. Toy designers, like many artistic individuals, have intuitive flashes of creativity. Often, that intuitive flash is, in reality, the product of earlier thought and practice in an artistic craft. We fail to see how the value of PlayWood's concept would differ in any respect had Clausi spent several months and several thousand dollars creating the noise-producing track. Accordingly, we conclude that PlayWood's lack of proof on this factor does not preclude the existence of a trade secret.

6. Ease or difficulty with which PlayWood's concept could have been properly acquired or duplicated by others

Finally, we also believe that there was sufficient evidence for the jury to determine that PlayWood's concept could not have been easily acquired or duplicated through proper means. PlayWood's expert witness, Michael Kennedy, testified: "This is a fairly simple product if you look at it. But the truth is that because it delivers feeling and sound as well as appearance, it isn't so simple [**41] as it first appears. It's a little more elegant, actually, than you might think." Trial Tr. at 504. In addition to Kennedy's testimony, the jury heard that Learning Curve had spent months attempting to differentiate its track from Brio's before Clausi disclosed PlayWood's concept of noise-producing track. From this evidence, the jury could have inferred that, if PlayWood's concept really was obvious, Learning Curve would have thought of it earlier.

Despite this evidence, the district court concluded that PlayWood's concept was not a trade secret because it could have been easily duplicated, stating that "had PlayWood succeeded in producing and marketing [the] notched track, the appearance of the track product itself would have fully revealed the concept PlayWood now claims as a secret." R.202 at 5-6. Of course, the district court was correct in one sense; PlayWood's own expert recognized that, in the absence of patent or copyright protection, the track could have been reverse engineered just by looking at it. *See* Trial Tr. at 562. However, the district court failed to appreciate the fact that PlayWood's concept was not publicly available. As Professor Milgrim states: "A potent [**42] distinction exists between a trade secret which *will be* disclosed if and when the product in which it is embodied is placed on sale, and a 'trade secret' embodied in a product which has been placed on sale, which product admits of discovery of the 'secret' upon inspection, analysis, or reverse engineering." 1 Roger M. Milgrim, *Milgrim on Trade Secrets* § 1.05[4], at 1-228 (2002). "Until disclosed by sale the trade secret should be entitled to protection." *Id.*; *see also* 2 Rudolf Callmann, *The Law of Unfair Competition, Trademarks and Monopolies* § 14.15, at 14-123 (4th ed. 2003) ("The fact that a secret is easy to duplicate after it becomes known does not militate against its being a trade [*730] secret prior to that time."). Reverse engineering can defeat a trade secret claim, but only if the product could have been properly acquired by others, as is the case when the product is publicly sold. Here, PlayWood disclosed its concept to Learning Curve (and Learning Curve alone) in the context of a confidential relationship; Learning Curve had no legal authority to reverse engineer the prototype that it received in confidence. *See Laff v. John O. Butler Co.*, 64 Ill. App. 3d 603, 381 N.E.2d 423, 433, 21 Ill. Dec. 314 (Ill. App. Ct. 1978) [**43] ("[A] trade secret is open to anyone, not bound by a confidential relationship or a contract with the secret's owner, who can discover the secret through lawful means."). Accordingly, we must conclude that the jury was entitled to determine that PlayWood's concept could not easily have been acquired or duplicated through proper means.

B. Exemplary Damages

The Illinois Trade Secrets Act authorizes exemplary damages of up to twice the amount of compensatory damages if there was a "willful and malicious misappropriation." 765 ILCS 1065/4(b). The jury was not given an instruction on exemplary damages because the district court granted Learning Curve's motion for judgment as a matter of law on this issue prior to closing argument. *See* Trial Tr. at 1355. PlayWood submits that the jury should have been permitted to determine whether Learning Curve's intentional misappropriation of PlayWood's trade secret in the realistic looking and sounding toy railroad track justified an award of exemplary damages. *See Medow v. Flavin*, 336 Ill. App. 3d 20, 782 N.E.2d 733, 746, 270 Ill. Dec. 174 (Ill. App. Ct. 2002) (stating generally that "the question of whether a defendant's [**44] conduct was sufficiently willful or wanton to justify the imposition of punitive damages is for the jury to decide") (quoting *Schmidt v. Ameritech Illinois,* 329 Ill. App. 3d 1020, 768 N.E.2d 303, 263 Ill. Dec. 543 (Ill. App. Ct. 2002)).

There are no Illinois cases construing the phrase "willful and malicious misappropriation" under the Act. However, we previously have held that the phrase includes "an intentional misappropriation as well as a misappropriation resulting from the conscious disregard of the rights of another." *Mangren Research & Dev. Corp. v. Nat'l Chem. Co., Inc.,* 87 F.3d 937, 946 (7th Cir. 1996); *see also Lucini Italia Co. v. Grappolini*, 2003 U.S. Dist. LEXIS 7134, No. 01 C 6405, 2003 WL 1989605, at *19 (N.D. Ill. Apr. 28, 2003); *RKI, Inc. v. Grimes,* 177 F. Supp. 2d 859, 879 (N.D. Ill. 2001); *Richardson Elecs., Ltd. v. Avnet, Inc.,* 1999 U.S. Dist. LEXIS 1138, No. 98 C 5095, 1999 WL 59976, at *5 (N.D. Ill. Feb. 6, 1999); *but see Roton Barrier, Inc. v. Stanley Works,* 79 F.3d 1112, 1120 (Fed. Cir. 1996) (holding that exemplary damages are not permitted under the Act when the defendant was motivated by competition, rather than [**45] by malice). n9

n9 In other contexts, Illinois courts routinely have held that

> "exemplary damages may be awarded when torts are committed with fraud, actual malice, deliberate violence or oppression, or when the defendant acts willfully, or with such gross negligence as to indicate a wanton disregard of the rights of others." *Kelsay v. Motorola, Inc.,* 74 Ill. 2d 172, 384 N.E.2d 353, 359, 23 Ill. Dec. 559 (Ill. 1978); *see also Medow v. Flavin,* 336 Ill. App. 3d 20, 782 N.E.2d 733, 747, 270 Ill. Dec. 174 (Ill. App. Ct. 2002); *Tucker v. Illinois Power Co.*, 232 Ill. App. 3d 15, 597 N.E.2d 220, 231, 173 Ill. Dec. 512 (Ill. App. Ct. 1992).

We agree with PlayWood that a rational jury could determine that exemplary damages are justified in this case. Specifically, we believe that a rational jury could determine that Learning Curve intentionally misappropriated PlayWood's trade secret in the noise-producing track and then attempted to conceal the misappropriation by creating [**46] false evidence of prior independent development. Accordingly, we remand this case to the district [*731] court with the instruction to hold a jury trial on exemplary damages. n10 We leave it to the district court on remand to consider PlayWood's request for attorneys' fees. *See* 765 ILCS 1065/5(iii) (permitting the court to award reasonable attorneys' fees to the prevailing party where "willful and malicious misappropriation exists").

> n10 We decline to consider PlayWood's argument that the district court erred by excluding under *Daubert v. Merrell Dow Pharmaceuticals, Inc.,* 509 U.S. 579, 125 L. Ed. 2d 469, 113 S. Ct. 2786 (1993), the testimony of Play-Wood's ink expert that a handwritten entry in Lee's "100-Day Agenda" relating to noise-producing track was not written contemporaneously with the other handwritten entries. Rule 10(b)(2) of the Federal Rules of Appellate Procedure provides that "if the appellant intends to urge on appeal that a finding or conclusion is unsupported by the evidence or is contrary to the evidence, the appellant must include in the record a transcript of all evidence relevant to that finding or conclusion." Fed. R. App. P. 10(b)(2). PlayWood did not request that the transcript of the *Daubert* hearing be included in the record on appeal. As a result, we are unable to

evaluate whether the district court erred in excluding the testimony of PlayWood's ink expert; accordingly, PlayWood has forfeited this argument. *See Hotaling v. Chubb Sovereign Life Ins. Co.,* 241 F.3d 572, 581 (7th Cir. 2001); *LaFollette v. Savage,* 63 F.3d 540, 545 (7th Cir. 1995); *Wilson v. Electro Marine Sys., Inc.,* 915 F.2d 1110, 1117 (7th Cir 1990). We recognize that, as an alternative to forfeiture, we have the authority under Rule 10(e) of the Federal Rules of Appellate Procedure to order PlayWood to supplement the record to include the *Daubert* hearing. *See* Fed. R. App. P. 10(e); *see also LaFollette,* 63 F.3d at 545. However, we decline to exercise that authority in this case because PlayWood "has had ample opportunity to correct the problem but has failed to do so." *LaFollette*, 63 F.3d at 545. Learning Curve pointed out in its answer brief that the *Daubert* hearing was not made part of the record on appeal. *See* Appellees' Br. at 41 ("PlayWood would have this Court undertake a *de novo* review and reverse the trial court's ruling on the admissibility of expert testimony, rendered after a *Daubert* hearing, without any reference to the evidence introduced at that hearing! Indeed, PlayWood has not included the hearing transcript in the record before the Court."). Despite notice of Learning Curve's objection to the incomplete record, PlayWood made no attempt to supplement the record or to explain why a transcript of the hearing was not necessary to permit meaningful appellate review.

[**47]

Conclusion

For the foregoing reasons, the judgment of the district court is reversed, and the jury's verdict is reinstated. The case is remanded to the district court for a jury trial on exemplary damages and for consideration of attorneys' fees by the court. PlayWood may recover its costs in this court.

REVERSED and REMANDED

Appendix C
Checklist Of Potential Trade Secrets

Research & Development

Algorithms
Analytical Data
Blueprints
Calculations
Chemical Processes
Compounds
Confidential Internal Identifiers for Information
Data Object Definitions
Data Object Models
Data Structures
Design Data and Design Manuals
Diagrams - All Types
Drawings - All Types
Engineering and Technical Specifications
Engineering Plans
Experiments and Experimental Data
Flow Charts
Formulas
Human-Machine Interface Prototypes
Inventions
Laboratory Notebooks
Lists of Components Needed to Make Products
Mathematical and Logical Formulas
Measurements
Mechanical Processes
Models
Physical Processes
Plans
Processes
Product Designs
Product Prototypes

Proprietary Equations
Proprietary Information Concerning Research and Development
Proprietary Technology Information
Prototypes
R & D Know-How and Negative Know-How (i.e., what does not work)
R & D Reports - All Types
Recipes
Research and Development Programs
Scientific Research and Development Procedures
Screen Design Prototypes
Software Development Methodologies
Software Prototypes
Source Code
Specifications for Quantitative Decision-Making Models
Structure Charts
Symbols
System Designs
System Interface Designs
Technical Plans
Test Records
Thermodynamic Processes
Vendor/Supplier Information
Working or Experimental Models of Products

Production/Process Information

Algorithms
Control Diagrams
Cost/Price Data
Designs
Engineering Plans
Flow Charts
Human Factors Engineering Information
Internal Components of manufacturing Systems
Internal Software Utilities
Lists of Components Needed to Make Products

Manufacturing Processes
Mean Time between Failure Analysis
Measurements
Models
Problem Resolution Procedures
Problem Resolution Records
Process/Manufacturing Technology
Product Life Cycle
Production Know-How & Negative Know-How
Proprietary Computer Run-Time Libraries
Proprietary Computer Software
Proprietary Devices and Machines
Proprietary Information Concerning Production/Processes
Quality Assurance Information
Quality Control Information
Quantitative Analyses
Quantitative Decision-Making Techniques
Software Development Methodologies
Software Engineering Information
Source Code
Special Production Machinery
Specification for Production Processes and Machinery
Specifications
Specifications for Quantitative Decision-Making Models
Statistical Analyses
Statistical Modeling Techniques
Statistical Models
Systems Integration Plans
Test Plans
Test Records
Tools

Vendor/Supplier Information

Cost/Price Data
Vendor and Supplier Information

Quality Control Information

Customer Service Methods
Environmental Analysis
Information Concerning Quality Control
Maintenance Data
Maintenance Instructions
Maintenance Know-How & Negative Know-How
Maintenance Procedures
Preventive Maintenance Methods
Problem Resolution Procedures
Prototyping Procedures
Quality Assurance and Control Methods
Quality Control Manuals
Quality Control Procedures
Quality Control Records
Quantitative Decision-Making Techniques
Software Development Methodologies
Software Development Procedures
Statistical Modeling Techniques
Technical Support Procedures
Test Plans
Test Records
Testing Methodologies
Testing Procedures
Tools
Training Methods
Troubleshooting and Debugging Formulas

Sales & Marketing Information

Competitive Analyses
Competitive Intelligence Information
Customer Development Cycle
Customer Information
Customer Needs and Buying Habits
Customer Service Procedures
Estimation Formulas
Focus Group Data
Internal Analysis of Current Economic Factors
Know-How Concerning the Management of Customer Confidence
Market Analyses
Market Survey Results
Marketing and Sales Promotion Plans
Marketing Plans
Marketing Techniques
Methods for Obtaining Greater Market Share
Product Life Cycle
Product Pricing Formulas and Methods
Proprietary Customer Lists
Proprietary Information Concerning Customers
Proprietary Information Concerning Sales & Marketing
Proprietary Sales and Marketing Studies and Reports
Sales Call Reports
Sales Forecasting Techniques
Sales Techniques
Strategic Alliance Information
Supply and Demand Models

Internal Financial Information

Balance Sheets
Bank Records
Budgets
Business Planning Techniques
Cash Flow Analysis

Cash Management Procedures
Cash Management Techniques
Computer Printouts
Cost/Benefit Analyses
Employee Compensation Systems
Enterprise Modeling Techniques
Estimation Formulas
Forecasts
General Financial Plans
Income and Expense Statements
Internal Analysis of Current Economic Factors
Internal Analysis of Projected Economic Factors
Internal Bookkeeping Records
Internal Financial Documents
Investment Records
Investment Strategies
Investment Techniques
Management Methods
Market Analyses
Operating Reports
Product Costs
Product Margins
Product Pricing Formulas and Methods
Profit and Loss Statements
Profit Plans
Proprietary Administrative Information
Proprietary Financial Information
Quantitative and Statistical Analyses
Sales Forecasts
Sales Quotas
Supply and Demand Models
Techniques for Analyzing Economic Factors

Internal Administrative Information

Business Plans
Cash management Procedures
Customer Codes
Decision Tables and Trees
Employee Compensation Programs
Employee Compensation Systems
Employee Incentive Programs
Employee Suggestion Programs
Flow Charts
Human Resources Policies
Implementation Plans
Industry Codes
Internal Analysis of Projected Economic Factors
Internal Bookkeeping Records
Internal Computer Software
Internal Management Policies
Internal Organization
Internal Organization Charts
Investment Programs
Investment Records
Investment Strategies
Key Decision Makers
Management Methods
Management Plans
Marketing Plans
Methods for Obtaining Greater Market Share
Part Numbers
Parts and Equipment Replacement Programs
Product Codes
Product Pricing Methods
Project Planning Systems
Proprietary Charts of Accounts for Accounting
Quality Assurance and Control Programs
Quantitative Decision-Making Techniques
Specifications for Quantitative Decision-Making Models

Statistical Modeling Techniques
Strategic Alliance Information
Strategic Business Plans
Strategic Legal Planning Techniques
Structure Charts
Succession Plan for Key Employees
Supply and Demand Models
Training Plans
Training Programs
Vendor and Supplier Codes
Vendor and Supplier Information

Appendix D
Trade Secret Audit Document Checklist

Access to all security policies.
Access to the process (documented or otherwise) of how policies and procedures are created, edited, and rescinded.
Asset management and ownership documentation.
BYOD/CYOD/COPE/COBO: Corporate policy on mobile devices.
Contractor non-disclosure and intellectual property agreements (all versions at all company locations).
Corporate policy handbook.
Corporate policy on public statements and presentations, including at trade shows, industry conferences, and sales calls.
Documented hiring processes surrounding initial and periodic background checks, interview processes, and job descriptions for employees with direct access to trade secret information.
Documents relating to computer access rights upon hire as well as changes to these computer access rights upon changes in job functions.
Documents relating to employee exit procedures, including exit interviews.
Employee and contractor training modules: trade secrets.
Employee computer and email retention policies
Employee handbook.
Employee non-disclosure and invention agreements (all versions at all company locations).
Employment contracts (all versions at all company locations).
Existing trade secret inventory.
Existing trade secret protection program: policies, procedures and practices.
Incident response plans.
One-way and two-way NDAs (all versions at all company locations).
Organization charts: executive team and departments.
Other electronic and other means to protect against the unauthorized acquisition, disclosure and use of non-public information.
Vendor selection, vetting, non-disclosure policies.

Appendix E
Sample Nondisclosure Agreement

WHEREAS, [YOUR NAME] agrees to furnish ____________________ certain confidential information relating to ideas, inventions or products for the purposes of determining an interest in developing, manufacturing, selling and/or joint venturing;

WHEREAS, ____________________ agrees to review, examine, inspect or obtain such confidential information only for the purposes described above, and to otherwise hold such information confidential pursuant to the terms of this Agreement.

BE IT KNOWN, that [YOUR NAME] has or shall furnish to ____________________ certain confidential information and may further allow ____________________ the right to discuss or interview representatives of [YOUR NAME] on the following conditions:

1. ____________________ agrees to hold confidential or proprietary information or trade secrets ("confidential information") in trust and confidence and agrees that it shall be used only for the contemplated purposes, shall not be used for any other purpose, or disclosed to any third party.

2. No copies will be made or retained of any written information or prototypes supplied without the permission of [YOUR NAME].

3. At the conclusion of any discussions, or upon demand by [YOUR NAME], all confidential information, including prototypes, written notes, photographs, sketches, models, memoranda or notes taken shall be returned to [YOUR NAME].

4. Confidential information shall not be disclosed to any employee, consultant or third party unless they agree to execute and be bound by the terms of this Agreement, and have been approved by [YOUR NAME].

5. This Agreement and its validity, construction and effect shall be governed by the laws of the State of Illinois.

AGREED AND ACCEPTED BY:

Witness: ________________

Name: _________________________________

Date: _________________

Name: _________________________________

[YOUR NAME]

Date: _________________

Appendix F
Sample Employee Trade Secrets Exit Interview Form

I acknowledge that the undersigned representative of the Patent Department of [company] has conducted an exit interview with me.

At this interview, my employment obligations to protect [company's] trade secrets and proprietary and confidential information were reviewed, including but not limited to the following types of information attached hereto as **EXHIBIT A**. I hereby acknowledge and warrant that if I have any doubt whether a particular item of information is considered to be a trade secret or confidential or proprietary information of [company], I agree to keep such information secret and not to disclose or use such information for my own benefit or the benefit of others without the express written consent of [company].

I further acknowledge and warrant that I have delivered to [company] all copies of any company documents and writings, research notes, data, sketches, analyses, financial, customer or sales information, strategic business plans, commercialization processes, product implementation plans, contracts, and the like, whether manually or electronically stored.

I also acknowledge and warrant that no copies, duplicates, sketches, uploads, downloads, or replicas of such information remain in my possession, custody or control.

Date: _______________

Witness: _______________________________

EMPLOYEE: ______________________________

Social Security Number: _____________________

[COMPANY]

By: ____

Appendix G
Trade Secret Examiner® System Features

General

- Manages trade secret metadata, not the actual trade secret, leaving the actual trade secret in its current, secure locations.
- Simple data entry screen permits general employees to enter trade secret metadata without specialized training or legal help.
- One-to-five scoring of the six factors of a trade secret.
- Classification using intuitive SFP (Subject, Format, Product) methodology.
- Specialized database developed for trade secret asset management.
- Two-stage system providing review of employee inputs, permitting acceptance, rejection, or editing of employee input data before adding trade secret metadata to the trade secret inventory.
- Retains audit trail of all additions and changes, including username and date/timestamp of all additions and changes. No data is ever deleted.
- Security analysis tools assist tailoring security measures to trade secret values, increasing security effectiveness while reducing costs.
- Litigation support tools assist attorneys in preparing for trade secrets litigation, reducing attorney hours and costs.
- Data integrity ensured through the use of court-accepted hashcode blockchaining.
- Five user permission levels - Entry, Analyst, Reviewer, Legal, Administrator -- allow restricting users to data and actions appropriate to their job assignment.
- Compatible with all state trade secret laws, the federal Defend Trade Secrets Act of 2016, and current international trade secret protections.
- Robust and scalable platform using Microsoft Visual C# and SQL Server.
- Available in standalone workstation, multi-user server, and cloud-based platforms.

Contained in Version 1

- Capture location information for each trade secret. Multiple locations possible per trade secret. Configure locations. Track location security.
- Hold captured info for review to either accept or reject the potential trade secrets.
- Inventory pages with sort functions.
- Edit pages to allow editing trade secret metadata.
- Analyze security page for finding security problems with trade secrets.
- Analyze history page for looking at historical data about a trade secret.
- Means for obsoleting a trade secret when it becomes generally known or is published.
- Means for printing any page or list.
- Division field for companies organized into divisions.
- Parameters of a trade secret: status, confidentiality level, negative know-how, patentable subject matter, combination.
- Means of configuring subjects, formats, products, locations, divisions, the six factors, status, confidentiality level.
- Means for managing user accounts and passwords.

Added in Version 2

- Internal blockchaining of all trade secret metadata to ensure data integrity and prove metadata contents and dates.
- Keyword assistance to aid customers in selecting keywords appropriate to a trade secret.
- Retention of all prior trade secret metadata reports for historical records.
- Parent/child relationships between trade secrets allows trade secret metadata to be gathered into combinations of trade secrets, or for large trade secrets to be decomposed into their component trade secrets.
- Parent/child relationships between products allows the definition of product lines and product groups, allowing trade secrets to be assigned to multiple products with a single entry.
- Capability to store open-format notes to trade secrets for archiving

additional information.

- Litigation codes allow trade secrets to be given aliases so they can be referenced in open court.
- Archiving of hash codes from trade secret source files to relate database entries back to the actual trade secret information.
- One-button report download of all metadata relating to a trade secret or trade secrets as of a specific date.

Added in Version 3

- Third-party disclosure tracking, including inbound and outbound trade secrets.
- Storing and archiving of NDA documents within the system. NDAs are hashed and block-chained for proof certain of contents and dates.
- Graphics package for graphing in multiple formats trade secret metadata within the database.
- Multi-user server and cloud-based capabilities.
- Multi-user collision detection and resolution.
- Bulk updates performed in background to enable users to continue without waiting for the bulk updates to complete.
- Assign specific products and product lines to an individual user, blocking that user's access to other products' and product lines' trade secret metadata.
- Assign specific subjects (e.g. Manufacturing, or Engineering) to an individual user, blocking that user's access to other subject's trade secret metadata.

Added in Version 4

- Automated, assisted, and manual elimination of redundancies in trade secret data resulting from multiple employees entering the same trade secret.
- Secure mobile access to trade secret entry via both iPad and smartphone.
- Importing of employee data from company payroll system.
- Storing and archiving of employee agreements within the system. Employee agreements are hashed and blockchained for proof

certain of contents and dates.

- Configuring and recording specific security measures applied to trade secrets on a company-wide, location-specific, and trade secret specific basis.
- IP address whitelisting specifies IP address ranges to which the Trade Secret Examiner server will respond. All login and page requests received from other IP addresses will be ignored, making trade secret metadata unavailable to the broader Internet.

Added in Version 5

- Two-stage blockchaining, allowing daily certification of database contents through a secondary blockchain operated on the Trade Secret Office® server. Independent third-party verification of the database without the necessity for continuous communications between the Trade Secret Examiner server and the Trade Secret Office server.
- Management report, as of any date, with charts and data about the trade secret database contents.

Appendix H
The Economic Espionage Act as amended by the Defend Trade Secrets Act of 2016:

Markup and Commentary
By Dennis Crouch
May 11, 2016 (Last updated on May 12, 2016)
(reprinted here with permission)

Contents

Introduction

Chapter 90 of Title 18 was originally codified via the Economic Espionage Act of 1996[1] and subsequently amended by the Foreign and Economic Espionage Penalty Enhancement Act of 2012.[2] These original provisions created criminal and civil causes of actions for acts of economic espionage and trade secret thefts. However, standing for civil actions was limited to the U.S. Attorney General – leaving the procedure little used.

The law has now been substantially amended through the Defend Trade Secrets Act of 2016 (DTSA) to create a private civil cause of action for trade secret misappropriation available to any injured party.[3] The new cause of action is designed as an additional layer of protection and is not intended to preempt any state law trade secret causes of action.

In addition to the new private cause of action, the DTSA law also provides a seizure mechanism available to a trade secret owner in order to preemptively prevent wrongful dissemination of a trade secret.

The following is a mark-up of the EEA showing changes made by the DTSA. A limited set of notes are provided for each section.

The amendments made by the DTSA are effective as of their enactment date, May 11, 2016 and the private cause of action for trade secret misappropriation applies to acts on or after that date. The employee immunity notice requirement applies to any contracts or agreements entered into or updated after May 11, 2016.

- Dennis Crouch, May 11, 2016

---

[1] Pub. L. 104–294, title I, § 101(a), Oct.11, 1996, 110 Stat. 3488.

[2] Pub. L. 112–269, § 2, Jan. 14, 2013, 126 Stat. 2442.

[3] Pub. L. ______ (May 11, 2016).

Amendments to Chapter 90 of Title 18

§ 1831. Economic espionage

Notes: Section 1831 was left unchanged by the DTSA. The section creates a criminal cause of action for economic espionage as defined below.

(a) In general.--Whoever, intending or knowing that the offense will benefit any foreign government, foreign instrumentality, or foreign agent, knowingly--

(1) steals, or without authorization appropriates, takes, carries away, or conceals, or by fraud, artifice, or deception obtains a trade secret;

(2) without authorization copies, duplicates, sketches, draws, photographs, downloads, uploads, alters, destroys, photocopies, replicates, transmits, delivers, sends, mails, communicates, or conveys a trade secret;

(3) receives, buys, or possesses a trade secret, knowing the same to have been stolen or appropriated, obtained, or converted without authorization;

(4) attempts to commit any offense described in any of paragraphs (1) through (3); or

(5) conspires with one or more other persons to commit any offense described in any of paragraphs (1) through (3), and one or more of such persons do any act to effect the object of the conspiracy,

shall, except as provided in subsection (b), be fined not more than $5,000,000 or imprisoned not more than 15 years, or both.

(b) Organizations.--Any organization that commits any offense described in subsection (a) shall be fined not more than the greater of $10,000,000 or 3 times the value of the stolen trade secret to the organization, including expenses for research and design and other costs of reproducing the trade secret that the organization has thereby avoided.

§ 1832. Theft of trade secrets

Notes: Section 1832 creates a criminal cause of action for theft of trade secrets. The DTSA amends the statute to increase the penalty against corporate organizations found guilty.

(a) Whoever, with intent to convert a trade secret, that is related to a product or service used in or intended for use in interstate or foreign commerce, to the economic benefit of anyone other than the owner thereof, and intending or knowing that the offense will, injure any owner of that trade secret, knowingly--

(1) steals, or without authorization appropriates, takes, carries away, or conceals, or by fraud, artifice, or deception obtains such information;

(2) without authorization copies, duplicates, sketches, draws, photographs, downloads, uploads, alters, destroys, photocopies, replicates, transmits, delivers, sends, mails, communicates, or conveys such information;

(3) receives, buys, or possesses such information, knowing the same to have been stolen or appropriated, obtained, or converted without authorization;

(4) attempts to commit any offense described in paragraphs (1) through (3); or

(5) conspires with one or more other persons to commit any offense described in paragraphs (1) through (3), and one or more of such persons do any act to effect the object of the conspiracy,

shall, except as provided in subsection (b), be fined under this title or imprisoned not more than 10 years, or both.

(b) Any organization that commits any offense described in subsection (a) shall be fined not more than ~~$5,000,000~~ the greater of $5,000,000 or 3 times the value of the stolen trade secret to the organization, including expenses for research and design and other costs of reproducing the trade secret that the organization has thereby avoided.

§ 1833. Exceptions to prohibitions

Notes: Section 1833 offers a set of immunities and exceptions to the causes of actions defined by the statute. The DTSA amendment spells out a whistleblower mechanism for individuals to disclose trade secret information to the government or in a court filing without creating liability under the statute.

The provision outlines a notice provision that employers can provide employees and contractors regarding the immunities. Employers who fail to provide the notice to an employee who is later sued for trade secret infringement will be prohibited from receiving enhanced damages or attorney fees.

For further analysis of this provision, see Dennis Crouch, *Immediate Action for Human Resource Departments on the Defend Trade Secrets Act*, Patently-O (May 10, 2016).

(a) IN GENERAL.—This chapter ~~This chapter~~ does not prohibit or create a private right of action for --

(1) any otherwise lawful activity conducted by a governmental entity of the United States, a State, or a political subdivision of a State; or

(2) ~~the reporting of a suspected violation of law to any governmental entity of the United States, a State, or a political subdivision of a State, if such entity has lawful authority with respect to that violation~~ the disclosure of a trade secret in accordance with subsection (b).

(b) IMMUNITY FROM LIABILITY FOR CONFIDENTIAL DISCLOSURE OF A TRADE SECRET TO THE GOVERNMENT OR IN A COURT FILING.—

(1) IMMUNITY.—An individual shall not be held criminally or civilly liable under any Federal or State trade secret law for the disclosure of a trade secret that—

(A) is made—

(i) in confidence to a Federal, State, or local government official, either directly or indirectly, or to an attorney; and

(ii) solely for the purpose of reporting or investigating a suspected violation of law; or

(B) is made in a complaint or other document filed in a lawsuit or other proceeding, if such filing is made under seal.

(2) USE OF TRADE SECRET INFORMATION IN ANTI-RETALIATION LAWSUIT.—An individual who files a lawsuit for retaliation by an employer for reporting a suspected violation of law may disclose the trade secret to the attorney of the individual and use the trade secret information in the court proceeding, if the individual—

(A) files any document containing the trade secret under seal; and

(B) does not disclose the trade secret, except pursuant to court order.

(3) NOTICE.—

(A) IN GENERAL.—An employer shall provide notice of the immunity set forth in this subsection in any contract or agreement with an employee that governs the use of a trade secret or other confidential information.

(B) POLICY DOCUMENT.—An employer shall be considered to be in compliance with the notice requirement in subparagraph (A) if the employer provides a cross-reference to a policy document provided to the employee that sets forth the employer's reporting policy for a suspected violation of law.

(C) NON-COMPLIANCE.—If an employer does not comply with the notice requirement in subparagraph (A), the employer may not be awarded exemplary damages or attorney fees under subparagraph (C) or (D) of section 1836(b)(3) in an action against an employee to whom notice was not provided.

(D) APPLICABILITY.—This paragraph shall apply to contracts and agreements that are entered into or updated after the date of enactment of this subsection.

(4) EMPLOYEE DEFINED.—For purposes of this subsection, the term 'employee' includes any individual performing work as a contractor or consultant for an employer.

(5) RULE OF CONSTRUCTION.—Except as expressly provided for under this subsection, nothing in this subsection shall be construed to authorize, or limit liability for, an act that is otherwise prohibited by law, such as the unlawful access of material by unauthorized means.

§ 1834. Criminal forfeiture

Notes: Section 1834 was left unchanged by the DTSA. The section links any "forfeiture, destruction, and restitution" orders to the requirements of 18 U.S.C. § 2323. Although the provision is titled "criminal forfeiture" the language of the provision itself appears would apply to the new private civil actions as well. Thus, any orders under Section 1836 for forfeiture, destruction, or restitution will be limited by this provision. For its part, Section 2323 provides for a set of procedures associated with forfeiture.

Forfeiture, destruction, and restitution relating to this chapter shall be subject to section 2323, to the extent provided in that section, in addition to any other similar remedies provided by law.

§ 1835. Orders to preserve confidentiality

Notes: Section 1835 includes provisions designed to better ensure that district courts treat trade secret disclosures in confidence. The newly added Section 1835(b) requires that district courts permit a trade secret holder file a brief under seal that explains why its trade secrets should be kept confidential by the court. Portions of the new provision are not expressly limited to actions under the EEA/DTSA and so may be interpreted to apply to all federal civil actions. If so, this section would become an oft-cited and powerful provision.

(a) IN GENERAL.—In any prosecution ~~In any prosecution~~ or other proceeding under this chapter, the court shall enter such orders and take such other action as may be necessary and appropriate to preserve the confidentiality of trade secrets, consistent with the requirements of the Federal Rules of Criminal and Civil Procedure, the Federal Rules of Evidence, and all other applicable laws. An interlocutory appeal by the United States shall lie from a decision or order of a district court authorizing or directing the disclosure of any trade secret.

(b) RIGHTS OF TRADE SECRET OWNERS.—The court may not authorize or direct the disclosure of any information the owner asserts to be a trade secret unless the court allows the owner the opportunity to file a submission under seal that describes the interest of the owner in keeping the information confidential. No submission under seal made under this subsection may be used in a prosecution under this chapter for any purpose other than those set forth in this section, or otherwise required by law. The provision of information relating to a trade secret to the United States or the court in connection with a prosecution under this chapter shall not constitute a waiver of trade secret protection, and the disclosure of information relating to a trade secret in connection with a prosecution under this chapter shall not constitute a waiver of trade secret protection unless the trade secret owner expressly consents to such waiver.

§ 1836. Civil proceedings ~~to enjoin violations~~

Notes: Section 1836 includes the new private civil cause of action for trade secret misappropriation and also provides for preliminary seizure of "property necessary to prevent the propagation or dissemination of the trade secret."

The provision defines remedies available, including damages and injunctive relief.

The three-year statute of limitations is provided that begins once the misappropriation is discovered our reasonably should have been discovered by the complainant. In a likely to be litigated provision, the limitations provision indicates that "a continuing misappropriation constitutes a single claim of misappropriation." This is designed in-part as a reaction to patent and copyright law where each infringing act is a new act of infringement with a new time-period for brining action (or collecting damages). Of course, misappropriation is often defined by the relationship between the parties. Whereas an employee may properly access trade secret files, that same action could be misappropriation once the individual is no longer an employee.

(a) The Attorney General may, in a civil action, obtain appropriate injunctive relief against any violation of this chapter.

~~(b) The district courts of the United States shall have exclusive original jurisdiction of civil actions under this section.~~

(b) Private Civil Actions.—

(1) IN GENERAL.—An owner of a trade secret that is misappropriated may bring a civil action under this subsection if the trade secret is related to a product or service used in, or intended for use in, interstate or foreign commerce.

(2) CIVIL SEIZURE.—

(A) IN GENERAL.—

(i) APPLICATION.—Based on an affidavit or verified complaint satisfying the requirements of this paragraph, the court may, upon ex parte application but only in extraordinary circumstances, issue an order providing for the seizure of property necessary to prevent the propagation or dissemination of the trade secret that is the subject of the action.

(ii) REQUIREMENTS FOR ISSUING ORDER.—The court may not grant an application under clause (i) unless the court finds that it clearly appears from specific facts that—

(I) an order issued pursuant to Rule 65 of the Federal Rules of Civil Procedure or another form of equitable relief would be inadequate to achieve the purpose of this paragraph because the party to which the order would be issued would evade, avoid, or otherwise not comply with such an order;

(II) an immediate and irreparable injury will occur if such seizure is not ordered;

(III) the harm to the applicant of denying the application outweighs the harm to the legitimate interests of the person against whom seizure would be ordered of granting the application and substantially outweighs the harm to any third parties who may be harmed by such seizure;

(IV) the applicant is likely to succeed in showing that—

(aa) the information is a trade secret; and

(bb) the person against whom seizure would be ordered—

(AA) misappropriated the trade secret of the

applicant by improper means; or

(BB) conspired to use improper means to misappropriate the trade secret of the applicant;

(V) the person against whom seizure would be ordered has actual possession of—

(aa) the trade secret; and

(bb) any property to be seized;

(VI) the application describes with reasonable particularity the matter to be seized and, to the extent reasonable under the circumstances, identifies the location where the matter is to be seized;

(VII) the person against whom seizure would be ordered, or persons acting in concert with such person, would destroy, move, hide, or otherwise make such matter inaccessible to the court, if the applicant were to proceed on notice to such person; and

(VIII) the applicant has not publicized the requested seizure.

(B) ELEMENTS OF ORDER.—If an order is issued under subparagraph (A), it shall—

(i) set forth findings of fact and conclusions of law required for the order;

(ii) provide for the narrowest seizure of property necessary to achieve the purpose of this paragraph and direct that the seizure be conducted in a manner that minimizes any interruption of the business operations of third parties and, to the extent possible, does not interrupt the legitimate business operations of the person accused of misappropriating the trade secret;

(iii)

(I) be accompanied by an order protecting the seized property from disclosure by prohibiting access by the applicant or the person against whom the order is directed, and prohibiting any copies, in whole or in part, of the seized property, to prevent undue damage to the party against whom the order has issued or others, until such parties have an opportunity to be heard in court; and

(II) provide that if access is granted by the court to the applicant or the person against whom the order is directed, the access shall be consistent with subparagraph (D);

(iv) provide guidance to the law enforcement officials executing the seizure that clearly delineates the scope of the authority of the officials, including—

(I) the hours during which the seizure may be executed; and

(II) whether force may be used to access locked areas;

(v) set a date for a hearing described in subparagraph (F) at the earliest possible time, and not later than 7 days after the order has issued, unless the party against whom the order is directed and others harmed by the order consent to another date for the hearing, except that a party against whom the order has issued or any person harmed by the order may move the court at any time to dissolve or modify the order after giving notice to the applicant who obtained the order; and

(vi) require the person obtaining the order to provide the security determined adequate by the court for the payment of the damages that any person may be entitled to recover as a result of a wrongful or excessive seizure or wrongful or excessive attempted seizure under this paragraph.

(C) PROTECTION FROM PUBLICITY.—The court shall take appropriate action to protect the person against whom an order under this paragraph is directed from publicity, by or at the behest of the person obtaining the order, about such order and any seizure under such order.

(D) MATERIALS IN CUSTODY OF COURT.—

(i) IN GENERAL.—Any materials seized under this paragraph shall be taken into the custody of the court. The court shall secure the seized material from physical and electronic access during the seizure and while in the custody of the court.

(ii) STORAGE MEDIUM.—If the seized material includes a storage medium, or if the seized material is stored on a storage medium, the court shall prohibit the medium from being connected to a network or the Internet without the consent of both parties, until the hearing required under subparagraph (B)(v) and described in subparagraph (F).

(iii) PROTECTION OF CONFIDENTIALITY.—The court shall take appropriate measures to protect the confidentiality of seized materials that are unrelated to the trade secret information ordered seized pursuant to this paragraph unless the person against whom the order is entered consents to disclosure of the material.

(iv) APPOINTMENT OF SPECIAL MASTER.—The court may appoint a special master to locate and isolate all misappropriated trade secret information and to facilitate the return of unrelated property and data to the person from whom the property was seized. The special master appointed by the court shall agree to be bound by a non-disclosure agreement approved by the court.

(E) SERVICE OF ORDER.—The court shall order that service of a copy of the order under this paragraph, and the submissions of the applicant to obtain the order, shall be made by a Federal law enforcement officer who, upon making service, shall carry out the seizure under the order. The court may allow State or local law enforcement officials to participate, but may not permit the applicant or any agent of the applicant to participate in the seizure. At the request of law enforcement officials, the court may allow a technical expert who is unaffiliated with the applicant and who is bound by a court-approved non-disclosure agreement to participate in the seizure if the court determines that the participation of the expert will aid the efficient execution of and minimize the burden of the seizure.

(F) SEIZURE HEARING.—

(i) DATE.—A court that issues a seizure order shall hold a hearing on the date set by the court under subparagraph (B)(v).

(ii) BURDEN OF PROOF.—At a hearing held under this subparagraph, the party who obtained the order under subparagraph (A) shall have the burden to prove the facts supporting the findings of fact and conclusions of law necessary to support the order. If the party fails to meet that burden, the seizure order shall be dissolved or modified appropriately.

(iii) DISSOLUTION OR MODIFICATION OF ORDER.—A party against whom the order has been issued or any person harmed by the order may move the court at any time to dissolve or modify the order after giving notice to the party who obtained the order.

(iv) DISCOVERY TIME LIMITS.—The court may make such orders modifying the time limits for discovery under the Federal Rules of Civil Procedure as may be necessary to prevent the frustration of the purposes of a hearing under this subparagraph.

(G) ACTION FOR DAMAGE CAUSED BY WRONGFUL SEIZURE.—A person who suffers damage by reason of a wrongful or excessive seizure under this paragraph has a cause of action against the applicant for the order under which such seizure was made, and shall be entitled to the same relief as is provided under section 34(d)(11) of the Trademark Act of

1946 (15 U.S.C. 1116(d)(11)). The security posted with the court under subparagraph (B)(vi) shall not limit the recovery of third parties for damages.

(H) MOTION FOR ENCRYPTION.—A party or a person who claims to have an interest in the subject matter seized may make a motion at any time, which may be heard ex parte, to encrypt any material seized or to be seized under this paragraph that is stored on a storage medium. The motion shall include, when possible, the desired encryption method.

(3) REMEDIES.—In a civil action brought under this subsection with respect to the misappropriation of a trade secret, a court may—

(A) grant an injunction—

(i) to prevent any actual or threatened misappropriation described in paragraph (1) on such terms as the court deems reasonable, provided the order does not—

(I) prevent a person from entering into an employment relationship, and that conditions placed on such employment shall be based on evidence of threatened misappropriation and not merely on the information the person knows; or

(II) otherwise conflict with an applicable State law prohibiting restraints on the practice of a lawful profession, trade, or business;

(ii) if determined appropriate by the court, requiring affirmative actions to be taken to protect the trade secret; and

(iii) in exceptional circumstances that render an injunction inequitable, that conditions future use of the trade secret upon payment of a reasonable royalty for no longer than the period of time for which such use could have been prohibited;

(B) award—

(i)(I) damages for actual loss caused by the misappropriation of the trade secret; and

(II) damages for any unjust enrichment caused by the misappropriation of the trade secret that is not addressed in computing damages for actual loss; or

(ii) in lieu of damages measured by any other methods, the damages caused by the misappropriation measured by imposition of liability for a reasonable royalty for the misappropriator's unauthorized disclosure or use of the trade secret;

(C) if the trade secret is willfully and maliciously misappropriated, award exemplary damages in an amount not more than 2 times the amount of the damages awarded under subparagraph (B); and

(D) if a claim of the misappropriation is made in bad faith, which may be established by circumstantial evidence, a motion to terminate an injunction is made or opposed in bad faith, or the trade secret was willfully and maliciously misappropriated, award reasonable attorney's fees to the prevailing party.

(c) JURISDICTION.—The district courts of the United States shall have original jurisdiction of civil actions brought under this section.

(d) PERIOD OF LIMITATIONS.—A civil action under subsection (b) may not be commenced later than 3 years after the date on which the misappropriation with respect to which the action would relate is discovered or by the exercise of reasonable diligence should have been discovered. For purposes of this subsection, a continuing misappropriation constitutes a single claim of misappropriation..

§ 1837. Applicability to conduct outside the United States

Notes: Section 1837 was left unchanged by the DTSA, but it offers important insight on how the law may be applied in the foreign context. Importantly, a U.S. corporation can be held liable for trade secret misappropriation under the DTSA regardless of whether the misappropriation occurred abroad.

This chapter also applies to conduct occurring outside the United States if—

(1) the offender is a natural person who is a citizen or permanent resident alien of the United States, or an organization organized under the laws of the United States or a State or political subdivision thereof; or

(2) an act in furtherance of the offense was committed in the United States.

§ 1838. Construction with other laws

Notes: Section 1838 was left largely unchanged – indicating that the EEA/DTSA does not preempt other trade secret laws. The one exception added by the DTSA is that the whistleblower immunity of Section 1833(b) preempts state law actions as well.

~~This chapter~~ <u>Except as provided in section 1833(b), this chapter</u> shall not be construed to preempt or displace any other remedies, whether civil or criminal, provided by United States Federal, State, commonwealth, possession, or territory law for the misappropriation of a trade secret, or to affect the otherwise lawful disclosure of information by any Government employee under section 552 of title 5 (commonly known as the Freedom of Information Act).

§ 1839. Definitions

Notes: Section 1839 includes definitions. The DTSA slightly amends the definition of trade secret from what was found in the EEA to bring the definition more in line with that of the Uniform Trade Secrets Act (UTSA) that was adopted by almost all U.S. states. A number of differences between the UTSA and DTSA remain. For instance, the UTSA includes the catch-all "information" that is not included within the DTSA.

The provision also defines "misappropriation" as used in Section 1836.

As used in this chapter--

(1) the term "foreign instrumentality" means any agency, bureau, ministry, component, institution, association, or any legal, commercial, or business organization, corporation, firm, or entity that is substantially owned, controlled, sponsored, commanded, managed, or dominated by a foreign government;

(2) the term "foreign agent" means any officer, employee, proxy, servant, delegate, or representative of a foreign government;

(3) the term "trade secret" means all forms and types of financial, business, scientific, technical, economic, or engineering information, including patterns, plans, compilations, program devices, formulas, designs, prototypes, methods, techniques, processes, procedures, programs, or codes, whether tangible or intangible, and whether or how stored, compiled, or memorialized physically, electronically, graphically, photographically, or in writing if--

(A) the owner thereof has taken reasonable measures to keep such information secret; and

(B) the information derives independent economic value, actual or potential, from not being generally known to, and not being readily ascertainable through proper means by, ~~the public~~ another person who can obtain economic value from the disclosure or use of the information; ~~and~~

(4) the term “owner”, with respect to a trade secret, means the person or entity in whom or in which rightful legal or equitable title to, or license in, the trade secret is reposed~~.~~;

(5) the term ‘misappropriation’ means—

(A) acquisition of a trade secret of another by a person who knows or has reason to know that the trade secret was acquired by improper means; or

(B) disclosure or use of a trade secret of another without express or implied consent by a person who—

(i) used improper means to acquire knowledge of the trade secret;

(ii) at the time of disclosure or use, knew or had reason to know that the knowledge of the trade secret was—

(I) derived from or through a person who had used improper means to acquire the trade secret;

(II) acquired under circumstances giving rise to a duty to maintain the secrecy of the trade secret or limit the use of the trade secret; or

(III) derived from or through a person who owed a duty to the person seeking relief to maintain the secrecy of the trade secret or limit the use of the trade secret; or

(iii) before a material change of the position of the person, knew or had reason to know that—

(I) the trade secret was a trade secret; and

(II) knowledge of the trade secret had been acquired by accident or mistake;

(6) the term ‘improper means’—

(A) includes theft, bribery, misrepresentation, breach or inducement of a breach of a duty to maintain secrecy, or espionage through electronic or other means; and

(B) does not include reverse engineering, independent derivation, or any other lawful means of acquisition; and

(7) the term 'Trademark Act of 1946' means the Act entitled 'An Act to provide for the registration and protection of trademarks used in commerce, to carry out the provisions of certain international conventions, and for other purposes, approved July 5, 1946 (15 U.S.C. 1051 et seq.) (commonly referred to as the 'Trademark Act of 1946' or the 'Lanham Act')'..

Additional Statutory Amendments

§ 1961. Definitions

Notes: Section 1961 is part of the racketeering statute and the amendment here adds economic espionage and theft of trade secrets to the list of predicate offences that may be considered "racketeering activity" under RICO. To be clear, the new chose for trade secret misappropriation is not added to the list.

As used in this chapter--

(1) "racketeering activity" means

(A) any act or threat involving murder, kidnapping, gambling, arson, robbery, bribery, extortion, dealing in obscene matter, or dealing in a controlled substance or listed chemical (as defined in section 102 of the Controlled Substances Act), which is chargeable under State law and punishable by imprisonment for more than one year;

(B) any act which is indictable under any of the following provisions of title 18, United States Code: . . . sections 1831 and 1832 (relating to economic espionage and theft of trade secrets), . . . sections 2314 and 2315 (relating to interstate transportation of stolen property), section 2318 (relating to trafficking in counterfeit labels for phonorecords, computer programs or computer program documentation or packaging and copies of motion pictures or other audiovisual works), section 2319 (relating to criminal infringement of a copyright), section 2319A (relating to unauthorized fixation of and trafficking in sound recordings and music videos of live musical performances), section 2320 (relating to trafficking in goods or services bearing counterfeit marks), section 2321 (relating to trafficking in certain motor vehicles or motor vehicle parts), sections 2341-2346 (relating to trafficking in contraband cigarettes), sections 2421-24 (relating to white slave traffic), . . .

(C) any act which is indictable under title 29, United States Code, section 186 (dealing with restrictions on payments and loans to labor organizations) or section 501(c) (relating to embezzlement from union funds),

(D) any offense involving fraud connected with a case under title 11 (except a case under section 157 of this title), fraud in the sale of securities, or the felonious manufacture, importation, receiving, concealment, buying, selling, or otherwise dealing in a controlled substance or listed chemical (as defined in section 102 of the Controlled Substances Act), punishable under any law of the United States, (E) any act which is indictable under the Currency and Foreign Transactions Reporting Act, (F) any act which is indictable under the Immigration and Nationality Act, section 274 (relating to bringing in and harboring certain aliens), section 277 (relating to aiding or assisting certain aliens to enter the United States), or section 278 (relating to importation of alien for immoral purpose) if the act indictable under such section of such Act was committed for the purpose of financial gain, or (G) any act that is indictable under any provision listed in section 2332b(g)(5)(B);

Non Codified Sections of the DTSA:

Section 2. Effective Dates and Rules of Construction

Notes: The DTSA was enacted on May 11, 2016 when signed into law by President Obama. Thus, under DTSA Section 2(e), the new misappropriation cause of action applies against any act that occurs on or after that date. Section 2(f) reiterates that the nothing in the DTSA is intended to preempt any other provision of law.

(e) EFFECTIVE DATE.—The amendments made by this section shall apply with respect to any misappropriation of a trade secret (as defined in section 1839 of title 18, United States Code, as amended by this section) for which any act occurs on or after the date of the enactment of this Act.

(f) RULE OF CONSTRUCTION.—Nothing in the amendments made by this section shall be construed to modify the rule of construction under section 1838 of title 18, United States Code, or to preempt any other provision of law.

(g) APPLICABILITY TO OTHER LAWS.—This section and the amendments made by this section shall not be construed to be a law pertaining to intellectual property for purposes of any other Act of Congress.

Section 4. Report on Theft of Trade Secrets Occurring Abroad.

Notes: The US Patent Office Director is required to provide a bi-annual report on trade secret thefts abroad along with recommendations.

(a) DEFINITIONS.—In this section:

(1) DIRECTOR.—The term Director means the Under Secretary of Commerce for Intellectual Property and Director of the United States Patent and Trademark Office.

(2) FOREIGN INSTRUMENTALITY, ETC.—The terms foreign instrumentality, foreign agent, and trade secret have the meanings given those terms in section 1839 of title 18, United States Code.

(3) STATE.—The term State includes the District of Columbia and any commonwealth, territory, or possession of the United States.

(4) UNITED STATES COMPANY.—The term United States company means an organization organized under the laws of the United States or a State or political subdivision thereof.

(b) REPORTS.—Not later than 1 year after the date of enactment of this Act, and biannually thereafter, the Attorney General, in consultation with the Intellectual Property Enforcement Coordinator, the Director, and the heads of other appropriate agencies, shall submit to the Committees on the Judiciary of the House of Representatives and the Senate, and make publicly available on the Web site of the Department of Justice and disseminate to the public through such other means as the Attorney General may identify, a report on the following:

(1) The scope and breadth of the theft of the trade secrets of United States companies occurring outside of the United States.

(2) The extent to which theft of trade secrets occurring outside of the United States is sponsored by foreign governments, foreign instrumentalities, or foreign agents.

(3) The threat posed by theft of trade secrets occurring outside of the United States.

(4) The ability and limitations of trade secret owners to prevent the misappropriation of trade secrets outside of the United States, to enforce any judgment against foreign entities for theft of trade secrets, and to prevent imports based on theft of trade secrets overseas.

(5) A breakdown of the trade secret protections afforded United States companies by each country that is a trading partner of the United States and enforcement efforts available and undertaken in each such country, including a list identifying specific countries where trade secret theft, laws, or enforcement is a significant problem for United States companies.

(6) Instances of the Federal Government working with foreign countries to investigate, arrest, and prosecute entities and individuals involved in the theft of trade secrets outside of the United States.

(7) Specific progress made under trade agreements and treaties, including any new remedies enacted by foreign countries, to protect against theft of trade secrets of United States companies outside of the United States.

(8) Recommendations of legislative and executive branch actions that may be undertaken to—

(A) reduce the threat of and economic impact caused by the theft of the trade secrets of United States companies occurring outside of the United States;

(B) educate United States companies regarding the threats to their trade secrets when taken outside of the United States;

(C) provide assistance to United States companies to reduce the risk of loss of their trade secrets when taken outside of the

United States; and

(D) provide a mechanism for United States companies to confidentially or anonymously report the theft of trade secrets occurring outside of the United States.

Section 5. Sense of Congress

Notes: Although non-statutory, courts at-times rely upon the sense of congress in interpretation of statutory provisions. The sense of Congress may also push the president to influence foreign governments to further protect trade secrets.

It is the sense of Congress that—

(1) trade secret theft occurs in the United States and around the world;

(2) trade secret theft, wherever it occurs, harms the companies that own the trade secrets and the employees of the companies;

(3) chapter 90 of title 18, United States Code (commonly known as the Economic Espionage Act of 1996), applies broadly to protect trade secrets from theft; and

(4) it is important when seizing information to balance the need to prevent or remedy misappropriation with the need to avoid interrupting the—

(A) business of third parties; and

(B) legitimate interests of the party accused of wrongdoing.

Section 6. Recommendation of Best Practices for Seizure Practice

Notes: The seizure provisions of the DTSA are one of the more controversial aspects of the legislation. The provision here requires the Federal Judicial Center to develop a set of best practices for the process

(a) IN GENERAL.—Not later than 2 years after the date of enactment of this Act, the Federal Judicial Center, using existing resources, shall develop recommended best practices for—

(1) the seizure of information and media storing the information; and

(2) the securing of the information and media once seized.

(b) UPDATES.—The Federal Judicial Center shall update the recommended best practices developed under subsection (a) from time to time.

(c) CONGRESSIONAL SUBMISSIONS.—The Federal Judicial Center shall provide a copy of the recommendations developed under subsection (a), and any updates made under subsection (b), to the—

(1) Committee on the Judiciary of the Senate; and

(2) Committee on the Judiciary of the House of Representatives.

Made in the USA
Monee, IL
04 November 2026

67387513R00203